EVA AMSEN

Hey, There's Science In This

Essays about science in unexpected places

First published by easternblot books 2024

Copyright © 2024 by Eva Amsen

All rights reserved. No part of this publication may be reproduced, stored or transmitted in any form or by any means, electronic, mechanical, photocopying, recording, scanning, or otherwise without written permission from the publisher. It is illegal to copy this book, post it to a website, or distribute it by any other means without permission.

Eva Amsen has no responsibility for the persistence or accuracy of URLs for external or third-party Internet Websites referred to in this publication and does not guarantee that any content on such Websites is, or will remain, accurate or appropriate.

Designations used by companies to distinguish their products are often claimed as trademarks. All brand names and product names used in this book and on its cover are trade names, service marks, trademarks and registered trademarks of their respective owners. The publishers and the book are not associated with any product or vendor mentioned in this book. None of the companies referenced within the book have endorsed the book.

Cover art elements by Lesia Hnatiuk used under Canva Free Content Licence. Author photo by Donna Ford.

First edition

ISBN: 978-1-7384934-2-5

This book was professionally typeset on Reedsy. Find out more at reedsy.com

Contents

Introduction	1
Rubber ducks and Lego	4
Scary music	7
Artist to peanut scientist	11
Lab test's hot spring origins	14
Of mice and men – a poem	18
Science of Stranger Things	22
A composer among chemists	27
Music genres	31
The zoo's many roles	35
Late night Japanese anime	40
Dalí and DNA	43
The Vatican Observatory	47
A bestiary of amazing animals	50
School of the Air	56
Museum parking garage	59
Manhattanhenge	62
Viral internet meme	65
Hiking trail geology	68
Songs about science	73
Painting with egg yolk	77
Eco homes from earth and mushrooms	79
Ballet turns	83
Can music help you study?	86
The board game Pandemic	91

Acknowledgements	94
Permissions	95
References	98
About the Author	126

Introduction

Hey, thanks for reading the introduction! You're about to read a collection of essays that all have an underlying theme of finding science in unexpected places. Over the course of this book, you'll visit a school in the Australian outback, a parking garage, the Vatican, downtown Manhattan, a few hiking trails and other places. You'll read about anime, a board game, a meme, ballet, Salvador Dalí and much more. It's truly all over the place, but there's at least a little bit of science in every chapter.

All these essays started out as blog posts or online pieces that I wrote between 2006 and 2023. I started blogging when I was a PhD student in biochemistry to satisfy my drive for curiosity about all the science that wasn't related to my daily research. I knew that I didn't want to continue to be a researcher. I loved science, but I wanted to learn a little bit about everything, not a lot about one little thing. Setting up my first science blog was a chance for me to write about science in art, science films, science in pop culture, science-themed travel and other surprising connections to science.

In the last two decades, many science blogs have come and gone – and I've written for far too many of them. Over the years I've blogged on my own website (easternblot.net), for *Nature Network*, as part of group blog *The Finch and Pea*, and on Medium. I also still write about art and science as a contributor to the Forbes website. You'll find blog posts and articles from all these places in this collection.

Some pieces in this book are personal reflections, others are more informative. One of the chapters has "found poetry" from scientific article titles that I've also performed live as a science comedy set. (Side note: while I generally use family-friendly language, I should probably let you know upfront that there is a mention of an adult theme in the title of one of the scientific articles in this poem and another one in the name of a Salvador Dalí painting. Blame them, not me.) To make sure that all these pieces formed a somewhat coherent collection, I almost entirely rewrote some of the original blog posts into chapters that would fit the overall theme of the book.

I also updated some of the content for modern times. The chapter about the board game Pandemic was originally written before we all lived through one in real life, and the chapter about a school that teaches entirely remotely needed a big overhaul for the same reason. Even the chapter about Yellowstone needed to be updated to mention the pandemic. (You'll find out why when you read it).

Over the course of editing this collection, I've also tried to smooth out some of the stylistic differences between pieces. The original posts were written for publications that used American or British spelling (or a confusing mix of both because my international background isn't partial to either). I decided to stick to Oxford spelling as a compromise. Oxford spelling uses most of the British conventions, such as "colour", "programme" and "analyse" but uses -ize in words such as "realize". Exceptions in this book include original titles of research papers or names of organizations that have been kept in their original American spelling where mentioned. For measurements, I used the metric system because this is a science book. However, the famous 2200-mile length of the Appalachian Trail is familiar to many hikers, so that also remains in the book (with the equivalent length in kilometres in parentheses).

Despite this attempt at moulding everything into the same style, some

INTRODUCTION

pieces still bear the unmistakable markings of a personal blog post while others are much more journalistic. Some chapters are ridiculous, some are thoughtful. The end result, I hope, is a short book of essays that captures the chaotic spirit of the heyday of science blogging in a more polished and less fleeting format.

Hopefully, you will learn something new from this book, even if it's just a fun fact to spice up your dinner conversations. You can tell people that *Stranger Things* sometimes includes real science, or that houses can be built from mushrooms and garbage, or that a hot spring in Yellowstone was crucial for modern genetics research. But mainly I hope that this book makes you say, *"Hey, there's science in this!"*

Rubber ducks and Lego

Hey, there's science in this flotsam

In January 1992, a container ship travelling from Hong Kong to Tacoma, Washington, hit a patch of bad weather in the middle of the Pacific Ocean. The boat bounced around on the high waves and was shaken about so much that twelve containers fell overboard. That's not unusual, because cargo ships lose a few hundred containers at sea every year. But when one of these containers opened during the accident, it spilled thousands of bath toys into the sea, ready for adventure.

The fleet of rubber duckies and other toys enjoyed their newly found freedom and bobbed along on the ocean waves, marking the start of an epic journey. After months at sea, some of the toys found their way to Sitka, Alaska, where a beachcomber found them washed up on a beach. This caught the attention of Curtis Ebbesmeyer, an oceanographer in Seattle with a penchant for unusual flotsam. A few years earlier, he had been intrigued by Nike sneakers that kept popping up on Washington beaches and he used his oceanography skills to track their journey back to a lost container at sea. Now, his attention was focused on rubber ducks. Over the next few months, more bath toys continued to wash ashore along the Alaskan coast so Ebbesmeyer had his work cut out for him. He used the dates and locations of the recovered plastic toys to figure out which ocean current had brought them there from the site of the storm at sea over 3,000 kilometres away.

Not all of the bath toys beached in Alaska. Some caught a current that turned them toward Russia, while others floated northward, through the Bering Strait. These unlucky ducks got stuck in Arctic sea ice some time after 1994, but while they were frozen in place, a very similar accident happened again. This time, a container with Lego figures fell off a cargo ship off the coast of England in 1997. These toys abandoned ship much closer to land than the rubber ducks did, so it didn't take very long for many of the Lego pieces to drift ashore in Cornwall, on the South coast of England. The colourful plastic figures stood out against the brown sand and other flotsam, so they were almost immediately spotted by regular visitors to the beach.

Meanwhile, many of the ducks were still stuck in Arctic ice. This could have been the end of their journey, were it not for climate change. After almost 10 years, the ice melted enough for some of the ducks to escape the Arctic region and several of them suddenly appeared again in 2003. The faded but intact toys landed on beaches on the East Coast of Canada, the Northern US and the Hebrides in Northern Scotland.

Even though the lost shipment was an accident, it inspired creative researchers to purposely use floating objects to study flow patterns. In 2008, NASA researcher Alberto Behar dropped 90 rubber ducks into holes in a glacier in Greenland to track the flow of melted glacial ice. His ducks were all labelled with his contact information in multiple languages, and Behar hoped that anyone who found them would contact him so that he could learn more about the ducks' journey in the glacier. The location of the found ducks would show where glacial water ended up, which would in turn help scientists better predict how the glaciers will move over time. Unfortunately, none of the ducks have been found. And even if one of these ducks washed ashore now, an email to Behar would sadly go unanswered since he died in 2015 when the small aircraft he was piloting crashed shortly after take-off near Los Angeles.

There's still a chance that Behar's ducks will eventually be found somewhere. After all, 25 years after the Lego container fell off its ship, beachcombers in South-West England still occasionally find Lego dragons among other flotsam in the sand. It's only a matter of time.

Scary music

Hey, there's science in this horror movie soundtrack

Scary movies are far less scary without their soundtrack. Editor Robert Ryang found a very funny way to show this when he entered a competition by the Association of Independent Creative Editors in 2005. This competition asked people to re-cut the trailer for a film so that it became a different genre. Ryang chose the trailer for *The Shining*. Using clips from the original movie trailer, he added a brand-new voice-over that made the 1980 horror film sound like a sappy and lighthearted family comedy. To top it all off, the trailer ends with the upbeat song "Solsbury Hill" by Peter Gabriel playing over the final clips of the trailer. The cheerful music completely changes the mood of the video.

So, music makes scary movies scary – but what makes scary music so scary in the first place? Daniel Blumstein thinks it has to do with animal distress calls. Blumstein is a professor in the department of Ecology and Evolutionary Biology at the University of California, Los Angeles. His research group has been studying the behaviour of marmots, and over the years Blumstein noticed certain patterns in these animals' distress calls. Once he started paying attention to it, he found these same patterns – called nonlinear noise patterns – in the alarm calls of other animals as well. But much to his surprise, Blumstein also noticed the same patterns in soundtracks of scary movies.

After picking apart the sounds found in music from different movie genres, Blumstein discovered that scary movies often included high-pitched screams with a lot of noise. In this context, "noise" doesn't mean "loud sounds", but it describes a disturbance in the sound. Think radio noise. According to Blumstein, these noisy screams in scary movies have the same effect on us as animal distress calls have on other animals in their social group. They evoke strong feelings, and we're evolutionary conditioned to find these sounds very disturbing.

Not all scary music is characterized by high-pitched shrieking. Sometimes you can create a spooky atmosphere simply by picking the right chords or notes. This works especially well with minor (sad sounding) chords and dissonant sounds. In the Middle Ages, one combination of notes was even known as the "devil's interval" because it sounds so uncomfortable. An interval in music theory describes a pair of notes with a certain distance between them. In the case of the devil's interval, it's equivalent to the distance between the notes A and E flat, for example, or between D flat and G. It's also called a tritone or augmented fourth interval. If you play both of the notes in this interval at the same time, they don't sound harmonious and collaborative but stressful and a bit strange.

The name devil's interval makes it sound mysterious and occult, but of course the name is an exaggeration. It's just a nickname because this odd chord didn't follow the conventional rules of music writing. It definitely sounds unpleasant, though, and it does a very good job of making scary music sound scary. You can find it in the opening violin chords of Saint-Saëns' spooky sounding "Danse Macabre", for example. Leonard Bernstein also uses it in *West Side Story* to symbolize the tension between the two rivalling gangs in the musical. But what makes it so uncomfortable to listen to? It all comes down to physics.

In any two-note chord, each individual note creates a sound wave that has its own distinct wavelength and frequency. When those two sound waves overlap with each other, the combination creates a new pattern. For many types of chords this produces a nice harmonic sound, but the two notes that make up the tritone or devil's interval have incompatible wavelengths. They never meet. Instead of creating a new regular pattern, they form a dissonant sound.

So, both high-pitched nonlinear noise and dissonant intervals create an uneasy feeling which makes them perfect for scary music. But that still doesn't cover all aspects of scary music. What about the low drone sound used in the background of many films to generate suspense? (Think *Jaws*, for example.) What about the use of nursery rhymes or singing children to juxtapose fear and innocence? Those sounds shouldn't be scary at all but we *made* them scary by putting them in horror films! We've learned to associate certain music with fear simply because it's used to accompany scary things on screen. For example, Beethoven's "Symphony No. 9" isn't inherently frightening, but many film buffs associate it immediately with horrifying violent scenes in *A Clockwork Orange*.

If you hear a piece of music often enough in a scary situation, you get used to linking that music to fear. Over the years, many film music composers have used the same four-note pattern to indicate that a threatening or scary moment was about to happen. The four notes themselves aren't musically scary, but we link them to fear from the context in which we usually hear them. Those four notes originally come from a 13th-century Gregorian requiem, the "Dies Irae". Ever since, many composers from Mozart to John Williams have used the same theme to indicate threat and terror, and it's now inextricably connected to scary scenes.

Association also explains why high-pitched screams and dissonant chords sound scary to us. There are scientific explanations for why they sound *unpleasant* to our ears – they remind us of distress sounds, or they don't form regular mathematical patterns – but unpleasant isn't the same as scary. We *made* these sounds scary by putting them in scary films. So when you next watch a scary movie, pay close attention to the soundtrack and listen for nonlinear high-pitched screams, devil's intervals, and all the other tropes of scary music. And if it gets too much? Just mute the sound: it won't be as scary any more without the soundtrack.

Artist to peanut scientist

Hey, there's science in this Chicago World's Fair artist's future

George Washington Carver was one of the most well-known African American scientists, who focused his agricultural research on new uses for peanuts. But despite being mostly associated with peanut science, Carver was also an artist. In fact, he didn't even consider science until he was already studying art and piano at university.

It wasn't easy for Carver to find a university that would accept black students like him in the 19th century, but he was welcome at Simpson College in Indianola, Iowa. During his college art classes he particularly loved and excelled at painting plants. His art teacher, Etta Budd, noticed his talent and encouraged Carver to enter his painting "Yucca and Cactus" in a local art exhibit. The painting was such a success at this exhibit that it was even selected as one of the artworks to represent Iowa at the World's Fair in Chicago in 1893, where it got an honourable mention.

Despite Carver's talents, Budd worried that he wouldn't be able to make a living as an artist. She wanted to help him out, so she suggested that he take his illustration skills to the botany department at Iowa State Agricultural College, where they always needed people who could draw the intricate details of plants as well as he did. Carver took her advice and switched his focus from art to learning all about plant biology. After receiving his bachelor's degree in 1894 and his master's shortly after in 1896, Carver took on a research position at Tuskegee Institute in Alabama. Now known as Tuskegee University, this is one of the historically black colleges that were opened in the United States after the abolition of slavery, during a time when black people were not allowed to attend other colleges. These days, Tuskegee's name is often mentioned in a negative context because of a controversial syphilis experiment that took place there in the mid-20th century, in which study participants were deliberately misled. However, this happened long after Carver's time there. When he worked at Tuskegee it was mainly known as a research institute where black scientists could focus on agricultural studies.

At Tuskegee, Carver's work was initially focused on helping farmers increase the yield of their crops. He didn't just do the research, but he also spent a lot of his time communicating with farmers to make sure that they knew the best ways to get the most out of their farms. Carver explained how fertilization and crop rotation restore nutrients to the soil, and the farmers took this knowledge on board to get better yields. Partly thanks to Carver's work, agricultural production in the South increased, but this led to a new problem. Now farmers were harvesting more crops than they could possibly sell and didn't know what to do with the surplus.

Carver thought he could find a solution for this as well. He redirected his research efforts to look for new uses for crops that would increase market demand and reduce the surplus. He found a lot. During his time at Tuskegee, Carver invented more than 100 new applications for sweet potatoes and over 300 different ways to use peanuts. Not all these methods were equally successful. For example, he came up with ideas for a shampoo and a face cream made from peanuts, which – like most of his inventions – were never produced commercially or even patented. But Carver's job was not to sell these products, just to figure out what *could* be done. His research in new uses for peanuts and his efforts in popularizing peanut crops have now become the work he is best known for, but throughout his career Carver never let go of his love for art.

One of Carver's peanut ideas was to use the peanut surplus to make new paints and dyes. He created dyes for leather, wood and textiles, but Carver also used some of his self-created paints for his own art. In 1941, *Time Magazine* published a piece to mark the opening of an exhibition of several of Carver's paintings at Tuskegee and mentioned that several of the paints he used in these pieces were made from vegetables and magnolias. And with that, Carver's career had come entirely full circle, from painting flowers for science to turning plants into paint.

Lab test's hot spring origins

Hey, there's science inside this hot spring

Yellowstone National Park is famous for its beautiful scenery and spectacular hot springs, but it also has a very special connection to modern laboratory biology. The park is the origin of a molecule called Taq polymerase, a compound which has been at the core of the fields of genetics, biochemistry, virology and many other molecular life sciences. It was even instrumental in many of the COVID-19 tests used in 2020 and 2021.

But how did all this modern science trace its roots to Yellowstone National Park? To find out, we need to go back to the 1960s, when biologist Thomas Brock did ecology research in the park to learn more about the colourful microbes that live in geysers and hot springs there. Brock hoped that studying the creatures in these hot water pools could help him understand how these microbes were able to survive and carry out photosynthesis under such extreme conditions.

Brock took samples from springs at different temperatures and found many more microorganisms than he imagined could possibly live in this environment. One of the locations he visited was Mushroom Spring, in the Lower Geyser Basin area of Yellowstone National Park. In autumn 1966, he collected a sample of hot water from this spring that included a new kind of microorganism that had never been seen before. Brock initially called his new species *Caldobacter trichogenes*, but soon changed the name to *Thermus aquaticus*, or Taq for short.

Together with all the other heat-loving bacteria that he found in Yellowstone, Brock dutifully recorded his sample and sent some of it to the American Type Culture Collection (ATCC). This facility has tubes full of all kinds of different cells – from microbe to mammal – that they make available to other scientists who want to use a particular kind of cell in their biology research. That is how a sample of Taq ended up in the labs of Cetus Corporation in California 17 years later. At Cetus, a scientist called Kary Mullis was working on a new method that could rapidly make lots of copies of small DNA samples. He needed the Taq sample for this method to work, but we'll get back to that in a moment. So let's put the Taq back in the ATCC freezer for now and jump ahead to see what Mullis was working on.

Mullis developed something called the polymerase chain reaction, or PCR for short. At the risk of sounding like an exaggeration, PCR has completely revolutionized any field in which experts study DNA. Any research involving genetics, cloning or identifying the function of new genes will at some point have involved PCR. It has also become a mainstay of forensics labs, and you might have seen PCR in action in the fictional crime labs of *Law and Order*, *CSI* and other crime shows and movies. It is everywhere.

It can also be used to study infectious diseases. If you were sick or travelling in 2020 or early 2021, you might have been asked to do a PCR test to check if you had COVID-19. In this case, you would have experienced the first part of the test, in which a sample of mucus was collected from your nose or throat (or both). This sample was then sent to a lab that has PCR machines and COVID-19 "primers" – small pieces of DNA that only matched genetic material from the SARS-CoV-2 coronavirus. If the virus was present in your sample, the PCR reaction would then use these primers to produce many small bits of DNA which are easy to detect as a marker of disease. It's an accurate method to test for infections, but later during the COVID-19 pandemic these tests were mostly replaced with rapid test strips that you use at home to get a much faster result. That's because the way that PCR works always takes a bit of time, and it can only be performed in a lab that has the right equipment.

PCR uses a series of temperature changes to untangle the two strands of a piece of DNA and uses each strand as a template to make more copies. If you imagine DNA looking like a ladder, picture the ladder being ripped in half, breaking every tread. Then recreate a ladder from each half by extending the broken treads and replacing the missing side. Now you have two ladders. It sounds tedious when it's a ladder, but for DNA this is quite a simple process. All it takes is repeated heating and cooling, and an enzyme called DNA polymerase that can form new strands of DNA on each half of the unravelled strand. Unravel those newly formed DNA strands, and the process starts over, each round doubling the amount of DNA.

When Kary Mullis was developing this new method in the early 1980s, one of the problems he needed to solve was that the high temperatures required to unravel DNA would also damage the DNA polymerase. All organisms have their own DNA polymerase, but our human version, for example, works best at our usual body temperature of about 37°C, and doesn't survive the high temperatures needed for PCR to work. Organisms that live at higher temperatures have polymerases that can better withstand heat, so Mullis and his colleagues just needed to find a polymerase from a creature that could survive at the high temperatures needed for PCR.

Now let's go back to that sample of *Thermus aquaticus* or Taq that Brock sent to ATCC in the 1960s. Grab it from the freezer and thaw it, because this was exactly what Mullis needed to make PCR a success. Since Taq is used to living in a hot spring, its enzymes – including its

DNA polymerase – have no problem at all with the extreme temperature swings that take place in a PCR sample.

Thanks to Taq polymerase, it's now possible to use PCR to rapidly replicate DNA samples for research in molecular biology, genetics, forensics, conservation, virology, epidemiology and much more. None of it would have been possible without the bacteria that Thomas Brock isolated decades earlier from a hot spring in Yellowstone. Taq polymerase was such a success, both scientifically and commercially, that in 2010 the United States' National Park Service introduced a Benefits-Sharing Environmental Impact Statement that asks researchers to sign an agreement before they collect samples and data on park grounds. The agreement states that, if research on park grounds leads to commercial success, the National Park Service will get a share of the profits. It was too late for them to get anything from the Taq discovery, though. Despite Yellowstone's crucial role in the process, the park never received any money from the molecule that Mullis brought to fame.

And Brock? He never got any money out of his discovery either, but he didn't mind. As quoted in a 2007 article in *Time* magazine, Brock said, "Yellowstone didn't get any money from it. I didn't get any money, either, and I'm not complaining. The Taq culture was provided for public research use, and it has given great benefit to mankind."

Of mice and men – a poem

Hey, there's science inspired by a Steinbeck novel title

Researchers who want to understand how human cells and genes work often look at mouse cells first. Why mice? They're surprisingly similar to humans. All mammals share roughly the same sets of genes, so if you need to study how a gene works in people you can also look at it in mice. Most countries heavily regulate how animals can be used for research, and researchers themselves are trying to use as few animals as possible (for example, by studying cells grown in dishes instead of using animals), but historically the mouse has been studied so often that it has become one of the default animal models that biologists use.

Because scientists know a lot about mice and their biology, it's no surprise that they also publish many articles in which mice are compared to humans. I often come across these articles when I'm researching a topic that I'm writing about, or if I am getting ready for a meeting with a scientist and want to understand their field a bit better. When I was still a researcher myself, most of the studies I read about were also done in mouse cells, so over the years I saw a lot of papers that compared mice and humans. At some point I noticed that many of these articles jokingly referenced John Steinbeck's novel *Of Mice and Men* in their title.

Being quoted in journal articles is by no means Steinbeck's first brush with science. He was friends with marine biologist Ed Ricketts and they even collaborated on the book *The Sea of Cortez,* based on a research expedition that Steinbeck joined in 1940. But their friendship has absolutely nothing to do with the reason that *Of Mice and Men* is referenced so often in the titles of scientific papers. That's purely because too many biologists all came up with the same joke for their research papers in which they compared mice and humans.

To highlight the absurdity of so many different research articles all quoting Steinbeck, I collected a few dozen of these *Of Mice and Men* paper titles and rearranged them into the poem below. Every line in the poem is the title of a real research article and you can find all 41 articles in the references section at the back of this book.

By the way, I once read this poem live on stage at a science comedy event. The Disney line got a good laugh, but of course this, too, is the title of a real and serious scientific publication!

Of mice and men?

Of mice and men.
Of mice and men, metals and mutations.
Of mice and men, corticosteroids, and vicarious participation.
Of mice and men – universality and breakdown of behavioral organization.
Of mice and men: the riddle of tubular regeneration.

Of mice and men: the human sciences and the humanities.
Of mice and men: skin cells, stem cells and ethical uncertainties.
Of mice and men.

Of mice and men, rats, and atherosclerosis.
Of mice, cats, and men: is human breast cancer a zoonosis?
Of mice and men.

Of mice and men: a model of HIV encephalitis.
Of mice and men: murine models of anti-GBM antibody nephritis.
Of mice and men: genetic skin diseases of keratin.
Of mice and men. Honesty and integrity in medicine.

Of mice and men.
Of digital mice and men.
Of cholesterol-free mice and men.
Of (stressed) mice and men.
Of (only) mice and men.
Of mice and men … but so much more too!

Of mice, men, and physicians.
Of mice, men, and trypanosomes.
Of mice, men, and cholesterol.
Of mice, men and the genome.

Of mice and men, and chandeliers.
Of mice and men … and elephants.
Of mice and men – and lopsided birds.
Of mice and men: the road to tolerance.

Of mice and men: the evolving phenotype of aromatase deficiency.
Of mice and men: an introduction to mouseology or, anal eroticism and Disney.

OF MICE AND MEN – A POEM

Of mice, and other beasts, and men.
Of mice and men (and cows and cats).
Of mice – and rats, dogs, rabbits, cats, and monkeys – and men.
Of fruit flies, mice, and men: the illicit review of science.
Of flies, mice, and men.
Of mice and men.

Of mice and men ... and broken hearts.
Of mice and men: the tale of two therapies.

Of mice and men: the mice were right.

Science of Stranger Things

Hey, there's science in this Netflix show

Much of the science in science fiction show *Stranger Things* is indeed fiction. You'll be pleased to know that there aren't really any secret labs full of children with mysterious powers and that there is no such thing as an upside-down alternate world loosely connected to ours. But that makes it even more exciting when real science occasionally does pop up in the show.

For example, in the third episode of Season 2, called "The Pollywog", Dustin visits his local library to try to find out as much as possible about a polliwog (or tadpole) that he found behind his house. While the creature in the show was stranger than any science book could possibly describe, Dustin – and the *Stranger Things* audience – learned some real biology facts in the process.

In the library, Dustin checks out five books, which are briefly shown in a shot of the library counter. One book has no visible title but a Dewey decimal number 597.9, which is a number within the reptile section of library books. I was able to read the title of two of the other books shown in this scene and those were relatively easy to find online: *Eastern Reptiles and Amphibians* by Roger Conant, and *The Book of Reptiles and Amphibians* by Michael H. Bevans. The Conant book is visible more clearly in a later scene where it's shown to be the

Roger Tory Peterson Field Guide edition from 1984 (the year in which *Stranger Things* Season 2 is set). Dustin got these books to look up more information on polliwogs (which the show spells as "pollywogs") after spotting something that looked like one in a garbage bin behind his house.

When Dustin first found his creature, it was in a stage of development where it had front legs, but no back legs yet. This is different from how most tadpoles develop. They usually get their hind legs first. But the most interesting feature about Dustin's pollywog, and what sets it aside from most amphibians, is that it doesn't live in water. Dustin kept it in a terrarium, where it thrived without any water, and named it Dart – short for d'Artagnan. If you've seen this season of the show, you'll know that the reason Dart doesn't need water is because it's a creature from the Upside Down, not catalogued by regular-world scientists. But before he realizes this, Dustin turns to his library books to try to find out if there are any amphibian species that don't need water in their early life stages.

There are indeed. Dustin tells his friends that neither the tadpoles of the *Indirana semipalmata* species nor those of *Adenomera andreae* need water. This is a bit of an anachronism, because the frog genus *Indirana* didn't get its name until 1986, while the episode is set in 1984. But other than that, the facts line up. These amphibian species, from India and South America respectively, really do have terrestrial polliwogs. They lay eggs on land and their tadpoles also grow up on land instead of in the water.

The writers and prop department of *Stranger Things* clearly went through considerable effort to add scientifically valid explanations about terrestrial polliwogs to the show, and they made sure to include real reference books that would include this kind of information. It might seem like a lot of unnecessary detail and work, but there is a good reason why producers and writers try to include real science in fictional shows and films wherever possible. They do this because having a link to real science makes the entire story seem ever so slightly more realistic. According to communication researchers, having a believable and realistic narrative makes the audience more engaged with the story. In this example, having Dustin study real amphibian science makes the audience realize that even the characters are aware how impossible Dart is. This makes it more relatable than if the characters immediately accepted the existence of a fictional monster.

If something is obviously wrong or impossible with no explanation, it distracts the viewer from the story, so the more *correct* the details are, the less distracting it is. For the same reason, screenwriters also sometimes use actual science as inspiration to create their fictional science scenes. There is a lot of very interesting scientific research that lends itself perfectly to fantastical stories. And from watching a bit more of *Stranger Things*, it looks like this is exactly what happened in Season 4. Without giving away too much of the main storyline, there is an element of music therapy in this season that isn't as far from reality as you might think.

In the fourth episode of *Stranger Things* Season 4, Robin and Nancy pose as psychology students on a visit to Pennhurst Mental Hospital. During their visit, they get a brief tour of the facilities and walk through a room where some of the patients are listening to music. The head of the hospital who shows them around says, "The right song, particularly one which holds some personal meaning, can prove a salient stimulus." And while most of *Stranger Things* is based firmly in fiction, this particular line clearly has roots in music therapy and neuroscience.

Music therapy is the practice of using music to treat mental or physical conditions. It has been around in some form since at least the 18th century, but it became much more established and professional in the 20th century, when hospitals in the United States began working with musicians to support veterans who returned from

World Wars I and II with post-traumatic stress disorder (PTSD). Besides managing PTSD, music therapy is now also regularly used to help people with depression, Alzheimer's Disease, Parkinson's Disease and other neurological conditions.

The effect of music on Parkinson's Disease is particularly striking because it changes physical behaviour. When people with Parkinson's Disease dance to music, they're temporarily better in control of their movements. The really fascinating thing about this is that this renewed control of balance and gait doesn't end when the music stops, but can last for several days after the dance session. Regular dance therapy even seems to slow down the rate at which Parkinson's symptoms get worse with time.

Scientists around the world have been studying the effect of music on Parkinson's for years. The research is even referenced in the 1990 film *Awakenings*, based on Oliver Sacks' book of the same name and on his discoveries (and those of others) made in the 1960s.

For Parkinson's dance therapy, it's all about rhythm. The beat of the music needs to be clearly heard, danceable, and within range of natural movements. But music therapy for Alzheimer's Disease works in a different way, and here the personal connection to the music is particularly important.

So what does all of this have to do with *Stranger Things*? Well, what Pennhurst's Warden Hatch describes is very similar to how music is used in Alzheimer's therapy. Scientists have confirmed that when people with early-stage Alzheimer's listen to songs that have close personal meaning to them, they use areas of the brain that are linked to cognition. Researchers also studied why listening to favourite songs seems to help people with Alzheimer's and found that it reduces stress and anxiety associated with the condition.

In the 1980s, during the time *Stranger Things* is set, the science behind music therapy wasn't as well understood as it is now. Even though a link between music and Parkinson's was known since at least the 1960s, it wasn't until 1989 that scientists first formally described the effect of music therapy for Parkinson's Disease.

But music therapy was already at the verge of becoming an established field back then. In 1983, the Certification Board for Music Therapists (CBMT) began formalizing the credentials of music therapists to ensure that the people who ran music therapy programmes didn't only have musical training but also knew how to handle a therapy setting. And in 1985, the World Federation of Music Therapy first got international music therapists together to discuss the field. That's the same year that Kate Bush released the song "Running Up That Hill" which also prominently features in Season 4 of *Stranger Things*.

It was also around the mid-1980s that researchers first started to realize and further explore how listening to music could affect people's brains and mental health. So it's not very far-fetched that the fictional Pennhurst Mental Hospital in *Stranger Things* would have been using music therapy. Robin and Nancy's visit to Pennhurst and what they learned there about music and the brain eventually drives the story forward and becomes very important throughout the season.

Storytelling experts would say that by basing some of this narrative in the real science of the time, an otherwise very fictional plot suddenly has a tiny aspect of believability in it, which further captures the audience. *Stranger Things* is far from the only show to include some genuine science to help drive the story. Many shows and film productions now work with science advisors and consultants to help them get the science as realistic as possible. Because sometimes, adding just a bit of factual science makes a fictional story more interesting.

A composer among chemists

Hey, there's science in this composer's life

The 19th-century Russian composer Alexander Borodin is not the world's best-known composer, but he has left an impressive legacy of music. His string quartets, symphonies and Prince Igor opera are still regularly performed around the world. But Borodin didn't see himself as a professional composer. He mainly composed for fun while his day job was in chemistry research. As a chemist, Borodin was close to some of the most important researchers of his time and one of his friends was Dmitri Mendeleev, who invented the periodic table which you'll find on the wall in many high school and university chemistry labs.

Borodin and Mendeleev met when they were both young researchers in the chemistry department of the University of Heidelberg. When Borodin arrived here in 1859, the institute was already home to some of the best chemists in the world. Pretty much anything that you might associate with that field of science today came out of this research institute. For example, if you picture one of those conical glass flasks whenever you think of chemistry, that was invented by Borodin's direct supervisor, Emil Erlenmeyer. And the department he worked in was run by Robert Bunsen, inventor of the Bunsen burner. But perhaps nothing is as iconic to chemistry as the periodic table that Dmitri Mendeleev would create in 1869.

The grid of the periodic table shows all the chemical elements sorted by size and common characteristics. The most impressive part of the table, and what makes it so important to chemists and physicists, is that it was able to predict the existence of certain elements before anyone had discovered them. For example, when Mendeleev sorted all known elements into his newly created table, he noticed a gap in between gallium (element 31) and arsenic (element 33). There should have been another element there, with somewhat similar properties to silicon and tin just above and below it in the table. Mendeleev was able to predict the approximate size and properties of an element that nobody had seen before. When the missing element, germanium, was indeed discovered in 1886, this emphasized that Mendeleev's table was an incredibly useful way to organize chemical elements. Modern chemists have been trying to complete the rest of the table by detecting every predicted element and, as of 2024, they have found 118 unique elements.

But before Mendeleev created his periodic table, he was a young postdoctoral researcher learning new methods to analyse different materials. Borodin was studying chemical synthesis in the same research institute as Mendeleev, but spent his spare time composing music. He worked on his "Piano Trio in D" around this time.

The two friends spent time together outside of work as well. In October 1860, Borodin and Mendeleev took a trip to Italy which briefly got Borodin arrested in a case of mistaken identity. The problem started because they were travelling light. "We wore only blouses, so that we would look like artists," Mendeleev once said about the trip, according to his biographers. "That's not a bad idea in Italy," he continued, "because you can get along very cheaply that way. We took hardly any shirts with us, and had to buy new ones when the need arose; we gave these away to the waiters in place of tips."

Looking nothing like the academics they were in Heidelberg, they reached the Kingdom of Lombardy-Venetia, which is now part of Northern Italy, but was still part of the Austrian Empire back then. While they were travelling through Lombardy-Venetia, Austrian police happened to be on the lookout for a political fugitive dressed in a bohemian outfit. In his loose-flowing travel outfit, Borodin vaguely matched the fugitive's description, so when officers spotted him, he was promptly taken in for questioning. Borodin had to explain to the Austrian border police that he was not at all the man they were looking for.

Borodin's actual life was probably much more interesting than that of the revolutionary fugitive the border police mistook him for. He was born in 1833 as the illegitimate son of Prince Luka Spanovich Gedianov, who entered his son's birth into the registry under the last name of one of his servants, Borodin. Even though his father did not legally accept Borodin as his son, he made sure that little Alexander and his mother had all they needed. That included music lessons, books and education to give him a good start in life. Borodin graduated from medical school in St Petersburg, and spent his free time making music. By the time he moved to Germany for his postdoctoral chemistry research, he had already composed several pieces for piano, voice, string ensembles and other instruments.

After his stint in Heidelberg, Borodin became a professor at the Medical–Surgical Academy in St Petersburg, and in the 1870s he launched the first university medical courses for women in Russia. He was never as well-known internationally for his chemistry research, especially compared to the high-profile chemists he spent time with in Heidelberg. But Borodin also had famous musician friends and is much better known for his musical life.

Once he returned to Russia after his research in Germany, Borodin worked closely with other Russian composers, including Modest Mussorgsky and Nikolai Rimsky-Korsakov. Together with César Cui and Mily Balakirev they formed a group called The Five, whose mission was to create a recognisable Russian style of music that was distinct from the classical music styles of Western Europe. Just listen to Rimsky-Korsakov's "Flight of the Bumblebee" or Mussorgsky's "Pictures at an Exhibition" and you'll get a sense of what they meant by that style.

Borodin's music is still popular with musicians and audiences around the world. He even posthumously won a Tony Award for composing the original score used in the musical *Kismet*, which was adapted from his music. But alongside his music career, Borodin always remained a scientist. He may not have made as much of a mark on chemistry history as his friend Mendeleev, but by all available accounts, Borodin was popular, likeable and always surrounded by music, up to the very moment of his death in 1887, when he collapsed with a heart attack while dancing at a costume party. What a way to go!

Music genres

Hey, there's science in music genres

Music genres are hard to define. They're partly related to factors of the music itself, such as its rhythm, melody, tempo, tonality or instrumentation. But it's more than that. Music genres are also based on historic and social concepts. When was the music written? Which country is it from? In what setting is it usually performed and how does the audience interact with it? Is this music to quietly listen to from the plush seats of a concert hall, or is this music to dance to at a crowded all-night rave in an abandoned warehouse?

Because there are conventions and norms associated with certain genres of music that aren't easy to identify from the music itself, two pieces of music that otherwise have a very similar musical structure could fall in completely different genres. For example, pop, rock and classical music all have similar rules about chord progressions – the way that chords follow each other within a piece of music. Some chord progressions are so persistently used throughout the ages and across different music styles that they have even become a bit of a running joke. A 2006 YouTube video by comedian Rob Paravonian shows him performing his "Pachelbel Rant", in which he points out music that uses the exact same chord progression as Pachelbel's "Canon in D" (a classical piece often played at weddings and ceremonies) across genres as diverse as folk, pop, rock and punk. A few years later, comedy group Axis of Awesome demonstrated something similar with their "Four Chord Song", which mashes up different songs that all use the same chords.

So when a group of Spanish computer scientists wrote an algorithm to sort music by chord progressions alone, it shouldn't be very surprising that they weren't able to clearly distinguish between pop music and classical music styles. Still, their algorithm could clearly pick up when something was jazz music, because jazz music has different rules for chord progressions than pop and classical.

In their study, the researchers used MIDI files to sort music by chord progressions. This file format includes a lot of information about the music, such as details about tempo, pitch, volume and other things that can be interpreted by computer programmes. But despite all that information, chord progression from MIDI files clearly wasn't enough to recognize different music genres.

Could there be another way for computers to sort music by genre? That's a challenge that computer scientists have been working on for a long time. They all start the same way, with a collection of MIDI files of musical pieces from different genres. Then they try to find patterns in these digital profiles that correspond to certain music styles.

In December 2017, a group of researchers in the United States studied how tonality evolved in Western music between the Middle Ages and the 20th century. Tonality describes which key a piece of music is in as well as chord progressions within the piece. To see how this characteristic changed over time, the researchers compared 24,000 songs from this period to look for patterns. For example, they found that most of the music in the set followed the traditional rules of Western music. There were some differences, though. The researchers noticed that certain patterns fell out of style over time or became more fashionable. They also found groups of patterns within the data that roughly corresponded with some known genres, but it wasn't perfect. One problem was that the system didn't consider any other factors besides tonality. This means that they grouped together composers from different eras that all happened to follow similar composition rules and patterns, but did not account for the fact that music genres are also determined by time periods. Even though you might find the same chords in 18th-century folk music and 20th-century chart toppers, the time period and context give you a hint that the second piece was probably pop or rock music. (Another side effect of only grouping by tonality was that notable atonal composer Schoenberg was all by himself in a group. Poor Schoenberg: even computers think his music is strange.)

So far, these studies suggested that looking at a single factor such as chord progression or tonality doesn't seem to be quite enough to define a music genre, but a few years ago, researchers in Venezuela found a different way to analyse MIDI files. They didn't just isolate the kind of information that a musician might understand (such as chord progressions), but they considered the entire collection of digital information that was stored in the MIDI file. They compressed this into the smallest possible set of symbols that still represented all the information of the piece. These symbols and patterns would not make things any clearer if a human looked at them, but computers could make sense of it. They looked at different ways that the information could be interpreted and created a profile of the diversity of the patterns they found.

When the researchers studied the profiles created by their algorithm, they noticed that most pieces in their collection had similar profiles to others within the same music genres. They could separate traditional music from pop music, or medieval music from more recent music, even if they used the same chords. Does this mean they could finally distinguish genres based on music data? Not quite. Even with this method, Baroque and rock music looked very similar, and Chinese music had similar profiles as Impressionistic music. The solution to understanding genres might lie in the information that's *not* in a MIDI file.

A MIDI file can't tell you which music award category a musician is usually nominated in, what festivals they perform at, which other music their fans listen to, how the musicians categorize themselves, or what subcultures their music is associated with. Music can be reduced to mathematical components, but we experience it as much more than that.

So to fully understand music genres, a computer needs more than MIDI files. For example, it could include behavioural data. Some music recommendation services such as Spotify already do this, by grouping music based on how many other users all listen to this same combination of musicians. Spotify doesn't even try to define or describe genres in the way that these music research studies have done. They just show how people interact with music. It's still not a flawless system, but it could bring computer algorithms another step closer to understanding what a genre is.

If computers do eventually learn how to pigeonhole music into genres, they will be one step ahead of humans, because we don't even really know how to unambiguously do this ourselves. Musicians, fans and music critics often have different ideas about which genre an artist or their work falls under, and they create new genres on the fly. There are subgenres, local genres and fusions, all determined by a combination of the music itself and the people who interact with it. Good luck figuring it out, computers.

The zoo's many roles

Hey, there's science behind our fun day out

When you think about zoos, you might be picturing a visit to the animals on a sunny spring weekend, eating ice cream while watching monkeys, that sort of thing. Well, your leisurely visit is only one side of the zoo. There's a lot more to zoos than you might think.

Originally, zoos were created as a place of entertainment where people could admire animals that were shipped in from far and wide. But over the years, the role of the zoo has shifted. Most zoos are no longer just about entertainment, but also about education, conservation and research. Together, these are the four main roles or purposes of a zoo.

Entertainment is just one of the four purposes of the zoo, but of course it's a big part of the appeal of zoos for many people. Visitors want a fun day out, and the zoo is there to provide that. It's also how zoos make money. Income from ticket sales helps to feed the animals and maintain their enclosures, so zoos are encouraged to advertise the entertainment angle of the zoo by telling people to come by to see the baby giraffes, meet a lemur, or gawk at the lions.

Not everyone enjoys the zoo. Many people make the point that we shouldn't keep animals trapped in enclosures for our own amusement.

Such ethical concerns are fair, and many countries have strict regulations about where and how animals can be shown to the public. However, for large zoos there are other reasons to have animals besides entertainment. One of these is education. It's still closely linked to entertainment because it relies on being able to show the animals to visitors. If you've ever been on a school trip to the zoo as a child (or as a teacher or parent), you'll have seen the education side of the zoo. Kids often end up learning about animals and meeting the staff during such trips, and some teachers add on a class discussion after the visit to go over what they learned. Of course, kids can learn about animals and biodiversity entirely from textbooks, videos and classroom lessons, but by visiting the zoo they get an experience that's much more engaging and interactive.

Education isn't just for school kids either. Even as a casual visitor to the zoo you'll probably end up learning something new from the displays near the enclosures or from "meet the zookeeper" sessions. Precisely how much learning happens at a zoo is difficult to measure, because you're likely to pick up information that you don't actively recall until much later. Still, researchers have found a few ways to track what people learned before or after a zoo visit. In a study at London Zoo, children were asked to draw animal habitats both before and after a visit to the zoo and to answer some basic questions. Overall, more than a third of children incorporated new things they learned at the zoo. However, the same study also noticed that if children didn't receive guidance from staff about the animals they saw, they sometimes left with even less understanding of conservation science than before. A much larger international study surveyed more than 5,000 zoo visitors of different ages and found that more people were able to correctly answer questions about biodiversity after they had visited a zoo than before. It wasn't a huge difference, but the participants were only tested before and after one zoo visit. There are other ways by which zoos encourage education: in the next chapter you'll read about a study that found that people are Googling the animals they see at their local zoo to learn more about them when they get home.

Besides entertainment and education, the zoo has two other purposes that aren't as visible to most visitors: research and species conservation. Simply put, we wouldn't know as much as we do about global wildlife if it weren't for the zoo. For example, in 2019, researchers tried to map how much we know about more than 30,000 species of mammals, birds, reptiles and amphibians from studying them in the wild. They discovered that scientists were still missing important information for over 98% of these species. Even biologists don't always know how animals survive and reproduce, which makes it difficult to plan conservation projects. But when researchers included data that was collected by scientists at zoos and aquariums, they suddenly had almost 800% more information than before! That's 800% of the small number of species they had enough information on in the first place, so it's still not very much (the number of well-understood species went from 613 to 4699), but it's a lot more than was available from field studies alone. Researchers who work at zoos are able to get very close to the animals, which makes it easier to study animal behaviour, diet and anatomy, and to carry out other research related to how animals live and survive. It doesn't always exactly reflect how animals would behave in the wild, but if there are no data available from wildlife observations, studies from the zoo are the next best thing.

One important area of research that zoos are involved in is conservation, with the purpose of increasing the population of endangered species. And it's working. If it weren't for zoo breeding programmes, several species would be a lot closer to extinction than they are now. Let's look at two animals to see how they were helped by zoo research. In the 1980s, both the California condor and the giant panda were endangered. The condor was the worst off, with only nine wild birds left in 1985. Around the same time, there were about 1,000 pandas still roaming around China, but there should have been many more. Researchers at zoos and national parks jumped into action to try to save both animals. One tactic that both the US condor researchers and the Chinese panda specialists tried was captive breeding.

In the 1980s, researchers at the U.S. Fish and Wildlife Service brought the condor into captivity so that they could try to breed them and get their numbers up. They captured all remaining wild condors and only bred them in zoos and research centres so that they knew exactly what was happening with their numbers and overall health. Starting in the 1990s, several of these condors were released and now there are more than 300 in the wild and over 200 in captivity.

The panda researchers had something similar in mind when they started a captive breeding programme, but the outcome here was very different. Unlike condors, pandas proved to be very difficult to breed in captivity and those that are born in zoos are unlikely to survive in the wild. But from studying the pandas they had in their breeding programme, researchers did get a much better sense of how to save them. Panda breeding projects have improved since then so that there are now a number of them in zoos around the world as a sort of backup panda plan (in case all the wild ones die) and as a living mascot for conservation. But captive breeding wasn't the only conservation action that China took. The other was to save the pandas' habitat by regrowing bamboo forests and creating space for the pandas to roam. That turned out to be the main answer to getting panda numbers back up. Even though panda breeding wasn't as successful as condor breeding, having pandas nearby in zoos helped researchers understand them better (even if that meant learning that they absolutely don't want to mate in captivity). That also supported their conservation because it made it extra clear that habitat restoration was the way to go.

So, zoos are a lot more complex than they seem, with roles in education, conservation and research as well as entertainment. Of course, there are still some badly run zoos that exist purely for people's amusement, but many other zoos have given us a lot of knowledge about animal conservation that we wouldn't otherwise have access to. And these efforts are at least partly funded by people who buy tickets to visit the zoo, so your fun day out could help researchers learn how to keep an endangered species in existence.

Late night Japanese anime

Hey, there's science in this anime

Ecologist Yuya Fukano planned to spend the summer of 2018 doing field work, but when an unexpected record-breaking heatwave hit Japan, he was forced to find a new research project that he could carry out from behind his desk instead of outdoors. Together with colleague Masashi Soga and zookeeper Yosuke Tanaka, Fukano used this time to explore the relationship between zoos and public interest in wildlife. The main question he was trying to answer was whether people wanted to learn more about the animals they saw at the zoo.

To study this, the researchers used publicly available data from Google Trends and Wikipedia views. This could tell them which animals people were looking up, when they were doing this search and in which cities these people were located.

One of the things they noticed was that the search volume for certain animals was higher in places where the local zoo showed these animals. For example, if a zoo in one city had hippopotamuses and another zoo didn't, they could see that difference by how many people were looking up hippopotamus information online in each city. That showed that people love to learn more about the animals they see at the zoo.

But Fukano noticed something peculiar in his data. In early 2017, the search volume for certain animals suddenly showed a massive increase. The average weekly search volume for some species more than doubled. What could make so many people suddenly want to learn more about the same animals all at once? The answer, of course, was television.

In January 2017, the anime *Kemono Friends* first aired on Japanese TV. This was a show about a girl who lives in a world where she is the only human surrounded by many anthropomorphized animals. By the time the first season of the show ended in March of that year, it had become so popular that rebroadcasts, merchandise, games and new seasons soon followed. Even though the show initially only aired in the middle of the night, Japan had become obsessed with *Kemono Friends*.

The *Kemono Friends* anime was based on an existing mobile game and manga from artist Mine Yoshizaki. The game, manga and show are set in a fictional zoo called Japari Park, in which all animals have been transformed to look like human girls while retaining some of their animal's original characteristics, such as ears or hair (fur) colour.

Some of the animal species in *Kemono Friends* are fictional (such as dragons) or extinct (like the dodo), but the majority of them are existing and currently living animals. In his research, Fukano noticed that 37 of *Kemono Friends*' many characters (26 mammals and 11 birds) did not naturally occur in Japan. So the only way people could get to know these 37 animals would be either from online research or by visiting the zoo. And that's exactly what happened. Immediately after the anime launched in January 2017, search results for these particular species went up. In the 18 months since *Kemono Friends*' first episode aired, people in Japan looked up these animals on Wikipedia over a million times more often than in the 18 months before that. Google searches were even higher: in those same months, the researchers recorded over 4.5 million additional Japanese language Google searches for the animals featured in *Kemono Friends* compared to the same period of time just before the show's launch.

The *Kemono Friends* media franchise also made it very easy for people to look up information about the animals. The show's official website listed information for each character that included their animal's scientific name and conservation status. For example, the information card for the character inspired by a leopard lists the name of the animal in Japanese and English as well as the official Latin name *Panthera pardus* and its status (vulnerable) on the IUCN Red List of Threatened Species. It also includes a description of the animal's habits and habitat. With so much scientific information on the show's official media channels, people were easily encouraged to learn even more and, as Fukano noticed, they started searching for information online.

But *Kemono Friends* didn't just bring people to search for information. They also inspired donations to wildlife conservation! Zoos in Tokyo ask their donors which animal they're supporting, and animals that were featured in episodes of *Kemono Friends* received more donations than other zoo animals after the show aired.

TV fame doesn't always come with such a positive impact, but in this case the show made a deliberate effort to incorporate wildlife education in their marketing and even had *Kemono Friends* characters displayed around zoos. The show's creators also donated to the World Wildlife Fund and to local zoos to support animal conservation.

Even though *Kemono Friends* already existed as a game and manga before the first episode of the show aired, it was really the anime that caused the big boost in online searches and zoo donations. To go from an anime broadcast at 1:30 in the morning to having fans actively searching for scientific information and supporting wildlife conservation is an amazing feat that shows the power that TV shows can have.

Dalí and DNA

Hey, there's science in this stereoscopic painting

Salvador Dalí is no stranger to unusual paintings. This is the man who painted droopy melting pocket watches in a desolate landscape, so nothing is too odd for this master of surrealism. But Dalí also painted the molecular structure of DNA. Now before you think that this sounds far too normal and average for a Dalí painting, let me quickly add that there is a pair of hands sticking out of the top of the DNA molecule in one of his paintings. Oh, and it can be viewed in 3D.

"The structure of DNA" is one of a pair of stereoscopic paintings that Dalí created in the 1970s. Each of these paintings exist as a matching set of two panels. At first glance, the left and right panels look similar, but the image is always ever so slightly tilted on the vertical axis. This makes it seem as if you're seeing the second panel from a different angle compared to the first panel. If the panels are placed directly next to each other and you view them from the right distance, the two images blend together and look like a single image that's coming out of the canvas towards you in the third dimension. For some of the images in Dalí's stereoscopic series, including "The structure of DNA", the left and right panel also have slightly different colours, which adds a further interesting effect when you see the combined image.

Creating images with viewing instructions so that you can see them in three dimensions is typical for Dalí's surreal style. But it's not a unique Dalí idea. For many decades, chemists and biochemists have been using this same method, called stereoscopy, to show the three-dimensional structure of molecules in two-dimensional print media such as textbooks or scientific journals. The more complex the molecules were that they needed to show, the more important it was to see them in 3D.

Most molecules are not flat but have atoms sticking out in all directions. This three-dimensional structure usually has an important function as well. For example, molecules that bind to receptors on the surface of a cell can act as medicine if they interact in a certain way with the three-dimensional binding site. These molecular structures are much too small to see with a microscope. Instead, they're often based on mathematical calculations of atomic and subatomic interactions that are measured with specialized equipment and translated into a three-dimensional structure. Some computer programmes show the 3D molecular model on screen and even let you rotate it and see it from different sides. It's easy to recreate something like that in a video or animation, but that doesn't help you much if the structure is only shown on paper. Especially back in the days when textbooks and scientific journals did not have online or digital supplements where such videos could be shown, the best way to convey the three-dimensional structure on a flat page was by using the same kind of stereoscopic images that Dalí painted.

When I was a chemistry student, one of my university textbooks included some of these stereo images of molecules. Just like Dalí's painting, they had a left and a right panel that looked almost identical but were slightly rotated compared to each other. The book even came with a plastic set of viewing glasses that you could assemble in such a way that they helped you view the structures. However, I quickly discovered that I could already easily see the 3D effect even without the glasses, thanks to the 1990s fad of Magic Eye pictures.

Magic Eye pictures look like colourful psychedelic patterns with no obvious image at first glance. But when you stare at them in just the right way, the patterns start to shift and reveal a hidden 3D image. These pictures were absolutely everywhere for a few years in the early 1990s. If you stared long enough at one of these posters you might see a boat or a bunny or a castle appear. The trick is identical to that of stereo viewing images with two panels. Instead of a left and a right panel, Magic Eye pictures have repeating patterns with slight variations. If you unfocus your eyes, the repeated sections of the pattern appear to move on top of each other, and the slight variations create the illusion that some elements are moving out of the page toward the viewer. Some people have an easier time seeing Magic Eye images than others. But if you can see them, it's equally easy to see stereoscopic molecule pictures or Dalí's series of two-panel stereoscopic images.

Dalí's stereoscopic paintings were some of his latest works, produced in the 1970s. But "The structure of DNA" was not the first time that he included DNA in his art. His first DNA painting was "Butterfly Landscape (The Great Masturbator in a Surrealist Landscape with DNA)". The strange subtitle refers to a recurring yellow figure in some of his work, always seen in profile, and shown in the sky above the DNA molecule in this painting. "Butterfly Landscape" was created in 1957, only four years after the structure of DNA was first discovered, and it is widely believed to be one of the very first DNA-inspired artworks, by any artist. In this painting, Dalí shows DNA in the helical ladder shape that most people are familiar with. The shape of the DNA molecule he painted for "The structure of DNA" is slightly different: it's essentially a detailed view of just one subunit of DNA (one side of one rung of the ladder, so to speak) where you can more clearly see how the individual atoms are organized in three dimensions.

Dalí included DNA in several other paintings over the years. And while he may have been the first artist to incorporate images of DNA structures in his work, he certainly wasn't the last. DNA continues to inspire artists, especially in the last few decades now that discussions about genetic sequencing or gene editing have put DNA into the public spotlight. But leave it to Dalí to get a textbook-style stereoscopic 3D view of the molecule into art galleries.

The Vatican Observatory

Hey, there's science at the Vatican

Astronomical observatories are usually built in large open spaces. You can find the Atacama Large Millimeter Array in the Chilean desert, the Mauna Kea Observatories on Hawaii's Big Island and the Roque de los Muchachos Observatory on La Palma in the Canary Islands. Observatories are in such remote places because being out in the middle of a desert or on an island makes it easier to see the sky without light pollution. So, the very last country you would expect to have an observatory is a country so small that it fits entirely within another country's capital city. That country is Vatican City – and it has an observatory.

To be fair, the Vatican's physical observatory isn't in Vatican City itself because of the aforementioned light pollution. You can't see space very well from the middle of Rome, where the Vatican is located. Instead, the Vatican has an observatory at the Pope's summer residence, Castel Gandolfo, just outside Rome, and another one at Mount Graham International Observatory in Arizona, which is also home to telescopes for other scientific organizations.

That the Vatican is doing astronomy research is not only surprising because of its size, but even more so because the Vatican is mainly known as the home of the Pope and the Catholic Church. However, it is exactly those religious ties that inspired the Vatican to study the night sky in the first place. Vatican astronomers have been studying the night sky since at least 1582. That year, Pope Gregory XIII decided that something needed to be done about the calendar, because it was clearly wrong. The equinoxes were not happening at the right calendar dates any more, which made religious holidays such as Easter end up on the wrong date. Solving this problem was a task for astronomers, who first had to figure out exactly how long a year should be. To make the calendars line up with the astronomical year again, the church decided that the year 1582 had to lose ten days, so the calendar that year jumped straight from October 4th to October 15th.

Even after the calendar issues were sorted out, there was still enough to do for the Vatican's astronomers. In 1891 Pope Leo XIII officially founded the Vatican Observatory to study the stars. It was initially located on Vatican grounds, but in the 1930s the city sky became too bright and a new observatory was opened at Castel Gandolfo. In just a few decades, light pollution from Rome and surrounding towns also started to affect the sky above Castel Gandolfo, so in the 1980s the second observatory site was set up in Arizona.

Vatican astronomers don't work in isolation. Just like other scientists they are involved in several large international collaborations. For example, in the late 19th century, the Vatican Observatory was one of the observatories that helped create a map of the sky, showing millions of stars that could be observed from telescopes at different sites on Earth.

The observatory still specializes in stars. One of its research focus areas is to understand how stars are formed and how they interact with the environment around them. Vatican astronomers are also interested in galaxies, cosmology and planets – both in our solar system and outside of it. The Vatican Observatory even joined the search for extraterrestrial life and has hosted meetings about astrobiology, the field that studies potential signs of biological life outside of Earth.

Astronomers from around the world also regularly visit the observatory's summer schools to learn more about a specific topic in their field. For example, in 2023 they discussed how to work with the extremely large datasets that are collected by telescopes. These datasets are impossible to sort through manually, so astronomers have to learn about data management and machine learning to analyse these large amounts of information, and the Vatican Observatory was one of the places where astronomers could learn more about this.

Behind all that research, the Vatican Observatory is still very much linked to the Vatican's main purposes as a religious site. The staff page of the observatory shows that most of the permanent staff at the facilities are Jesuits and priests. In 2019, Vox published an interview with Guy Consolmagno, who was a planetary scientist for years before he became a Jesuit brother and eventually landed a position as director of the Vatican Observatory. Interviewer Byrd Pinkerton asked Consolmagno to comment on the relationship between science and religion. He answered: "My religion tells me God made the universe. My science tells me how he did it."

Even though the motivation and origins of the Vatican Observatory are different than that of other observatories, they're looking at the same sky as everyone else and work on the same problems that other astronomers are studying. The only reason that the existence of the Vatican Observatory is surprising at all is because religion and science are often pitted as polar opposites. The Vatican itself has an unfortunate history with this, when in 1633 the Roman Catholic Church forced Galileo to deny that the Earth moves around the Sun. (They finally apologized three centuries later.) But part of the reason that the Vatican has actively been involved in astronomy since the 18th century is to show that the church doesn't have a problem with science. And they shouldn't. After all, science is for everyone, including the Vatican.

A bestiary of amazing animals

Hey, there's science in these fantastical medieval books – well, a little bit

In Medieval Europe, if you wanted to learn anything about animals you'd never seen before, your best bet would be a bestiary. These were books that included descriptions of amazing animals such as lions, elephants or exotic birds. But don't expect to learn many scientific facts from a bestiary. The animal descriptions in these books were usually allegories that reflected Christian symbolism linked to these creatures. They were not expected to be taken as scientific fact, and there was no scrutinous fact checking. If someone said that they once heard that elephants have no knees, then sure, why not put that in the bestiary? These inaccurate descriptions would then often be copied into other books as well, so they had a tendency to become quite persistent ideas.

Bestiaries did not distinguish between existing and imaginary animals. Both lions and dragons could be used as examples to discuss religious morals and values, so real and mythical beasts were presented side by side. For example, several 13th-century bestiaries have pictures and descriptions of the manticore, an imaginary animal with the head of a man, body of a lion and a scorpion's tail. They also included warnings about the basilisk, which supposedly ruled over all snakes and could kill people by meeting their gaze.

Towards the end of the Middle Ages and into the Enlightenment period, people became more interested in the natural sciences and started using books to share factual knowledge as well. Encyclopaedias, first developed in ancient Greek and Roman times, became popular again. Original bestiary descriptions of some animals were included in some of these early encyclopaedias and only updated later, when more information became available. The origins of modern encyclopaedia-style collections of animal facts with short descriptions and pictures, such as the IUCN Red List of threatened species, still has some resemblance to old bestiaries, even though the content is obviously much more factual and scientific!

Bestiaries also indirectly inspired fictional animal encyclopaedias, such as the Pokédex that collects facts about all Pokémon, or collections of mythical creatures in Dungeons and Dragons lore. In the Harry Potter series of books and films, a book called *Fantastic Beasts and Where to Find Them* plays a central role. This book is very strongly inspired by medieval bestiaries, with its description of mythical creatures, and it was also published as a standalone book and inspired its own series of spin-off films. When the first of these films came out on DVD in 2017, I wrote a blog post called "Seven real-life fantastic beasts and where to find them", in which I – much like Enlightenment-era natural historians – used the book of mythical beasts as a template to list facts about real animals like the platypus or the tardigrade. So forget about mythical creatures and any lessons in morality for this mini bestiary of some of the most fantastic beasts that you can find on Earth.

Axolotl

Why they're fantastic: Axolotls are a type of salamander that always stays in the larval stage. Most amphibians start out in water and then undergo metamorphosis to adapt to land. Not the axolotl! It can grow to adulthood without ever leaving the larval stage. Axolotls can also regrow limbs that they lost, and this superpower has made them an interesting subject for scientists who study regenerative biology and stem cell science. They want to find out the basic biology that explains what makes axolotls so good at growing new organs when other animals can only do that in their embryonic stage.

Where they live: Outside of labs and home aquariums, the axolotl is critically endangered in the wild. Axolotls are originally from Mexico, and there are still a few living in lakes over there.

Platypus

Why they're fantastic: The platypus is one of only five egg-laying mammals in the world. (The other four are all different species of echidnas.) When explorers first brought back dead platypuses from Australia to Europe, people thought they were fake because they looked like small beavers with a duck bill. But no, they're real, and they're fantastic. Platypuses lay eggs, but when their babies hatch they nurse them in the same way that other mammals would.

Where they live: Platypuses can be found swimming in ponds and streams in Eastern Australia.

Coelacanth

Why they're fantastic: What if an animal you only know from ancient fossils was still alive today? It sounds impossible, but this is exactly what happened to the coelacanth. When the first living coelacanth was caught in South Africa in 1938, museum worker Marjorie Courtenay Latimer recognized it as something special and had it stuffed to preserve it. A few months later, fish researcher J.L.B. Smith immediately identified it as a coelacanth, a fish he'd only seen in fossils from the Cretaceous period. Even after millions of years, the coelacanth was still recognizable, but hadn't stayed entirely unchanged. When researchers studied the coelacanth's genetic material, they learned that it is still evolving, just very slowly! One of the other quirks of the coelacanth is that its eggs hatch inside the mother so that the baby fish are birthed alive instead of in eggs.

Where they live: Coelacanths live in underwater caves along volcanic coastlines in East Africa and Indonesia.

Aye-aye

Why they're fantastic: The aye-aye might be one of the strangest primates. It's nocturnal, with enormous eyes, and its mouth looks a bit like that of a rabbit, with continuously growing front teeth. They have one extremely long and creepy middle finger, which they use to pull grubs out of trees after luring them by tapping on the tree, much like a woodpecker would do (but with a finger).

Where they live: The aye-aye is from Madagascar, where lots of unique primates live.

Tardigrade

Why they're fantastic: This tiny creature has eight legs, looks like a microscopic teddy bear, and can survive the most extreme conditions. They can be frozen for years at very low temperatures without dying. They can go 30 years without food. They're essentially immune to radiation and they can survive in space. Some of them can also tolerate high temperatures, but over the long term tardigrades do prefer it cool. They don't do well if the temperature is consistently warm, so climate change may be a problem even for tardigrades with all their superpowers.

Where they live: Like the axolotl, the tardigrade's fantastic properties have made it an interesting research subject for scientists, so the most likely place to actually see one is under a lab microscope. But in the wild, they're in lakes, ponds, meadows, soil, moss and many other places all over the world. They're basically everywhere and they're probably in your garden right now. Go say hi!

Giant Pacific Octopus

Why they're fantastic: It's huge and looks like something straight out of someone's imagination, but the giant Pacific octopus is real and it is awesome. Just like other octopuses, it can squeeze its body through tiny openings and is a genius at figuring out puzzles. That combination has made the octopus an excellent escape artist. The giant Pacific octopus can also change the colour of its skin to camouflage itself while it hunts, and it has not one, not two, but three hearts.

Where they live: As the name implies, the giant Pacific octopus lives in the Pacific Ocean. It's mainly found near the coast in the Northern Hemisphere.

Pangolin

Why they're fantastic: Pangolins are the only known mammals with scales. They're covered head to toe in their own body armour. Their scales are made of keratin, the same material that hair and nails are made of. Despite this heavy protective layer, pangolins are unfortunately endangered. Their biggest threat is humans, who hunt them for their flesh and scales even though the scales look best on a living pangolin.

Where they live: There are a few different species of pangolin, and they can be found in Asia and Africa.

School of the Air

Hey, there's science in remote learning

In the summer of 2009, I visited Australia for a few weeks. The trip was meant to be a well-deserved break after I finished my PhD, but I clearly couldn't keep away from an education environment, because during my travels I visited a school in the desert town of Alice Springs.

It was August, so the Australian school year was in full swing on the day that I locked my rented bicycle outside the school building and stepped inside. But even though school was in session, there were no students to be found anywhere. The school grounds were empty. Hallways were deserted. The kids were all at home, because the school I visited is one of the world's oldest long-distance education programmes, the School of the Air.

The school was founded in 1951 to provide education to children living in the Australian outback. Families in these remote communities were supplied with radio equipment and mailbags so that their children could connect with teachers and classmates who lived up to 1,000 kilometres away. Fifty years later, the school started using the internet, and now the kids log in to school software and see their teacher on video. After the COVID-19 pandemic, this doesn't sound unusual, but these kids were learning like this decades before the rest of the world caught up! When I visited the School of the Air, the iPhone was only two years old and very few people had smartphones. Broadband was slow and Wi-Fi was not as ubiquitous as it is now. Teaching online was simply not feasible for most schools, but the School of the Air was doing it. They were using a satellite connection and custom-built software so that students could interact and see both the teacher and a screen with a desk or whiteboard at the same time. That was a big step up from only hearing each other over a radio connection.

I saw the software in action during a public tour of the school. In one of the classrooms, which functions as a studio, a teacher was presenting a class to first-grade students. A camera captured the surface of her desk so that the kids could see what she was doing. On a screen beside her we could see the lesson in progress from the pupils' point of view, and over the speakers we heard a handful of six-year-olds answering the teacher's questions:

"How many groups of animals are there?"

"Two groups!"

"And how many in each group?"

"Three!"

This is everyday learning for many young people in the outback. A guide from the school's visitor centre explained that, at the time, 171 students in an area the size of Central Europe logged on to their classes at their assigned times, but that they never had more than four hours of online classes per week. The rest of the time they studied at home with a tutor who is familiar with the school's curriculum and the software. These tutors are usually parents, but some families hire a "govie" (short for governess), often a recent high school graduate who spends a year tutoring before heading off to university.

For science classes, which are often very interactive, the teacher usually demonstrates the experiment on screen so that students can try it at home with their tutor. A display at the school showed notes from a girl who built a sundial at home, and in the main area of the visitor centre was a model of an energy-efficient house created by a grade eight student as a science project. The school makes sure that students have everything they need to complete their lessons, whether that's online material or things sent to their home.

When School of the Air students graduate, at around age 15, they often leave for boarding school to complete their high school education. For many of them, that will be their first experience in a regular classroom setting. At the School of the Air, students only see their classmates in person a few times a year, when they travel hundreds of kilometres to attend "in-school weeks" to meet up with their friends and teachers.

Eleven years after my visit to the School of the Air, the COVID-19 pandemic forced many schools to quickly transition to online education. Since then, education researchers have looked into the effect that this global switch to remote learning had. Several studies found that students often did worse at standardized tests than they had before the pandemic. Parents and teachers struggled as well, because they couldn't always handle the additional demands of having kids learning from home. But researchers also noticed that about a third of students had no learning loss. What seemed to make a difference for these students was support at home, access to technology, quality of remote resources and plenty of feedback from teachers.

The Australian government also published a report with recommendations for remote learning, based partly on experiences with the School of the Air, and they came to the same conclusions: to make remote learning work, schools and homes need to be very prepared and have access to technology as well as support. These things weren't easy to come by for schools who had to abruptly change to online learning. Some managed better than others, but none were likely as well-prepared as the School of the Air.

Museum parking garage

Hey, there's science under this parking garage

S cience news is always a little bit behind. Updates about sports, politics, wars and natural disasters make their way into the spotlight on the very day that their newsworthy events happen, but science usually doesn't become news until months or years after the initial exciting discovery was made. For example, when the *New York Times* reported in February 2009 that a mammoth was found underneath a Los Angeles parking garage, that was news because the Page Museum at the La Brea Tar Pits finally officially announced that it was studying this mammoth skeleton. But they'd actually been researching it for several months already. It wasn't news to me, because I had seen the mammoth tusks half a year before the news broke, and so did everyone else who visited the tar pits during that time.

The La Brea Tar Pits might just be the most interesting place in Los Angeles. It's a place where crude oil has been coming up to the surface for at least 40,000 years, unbothered by the city that was built around it. I first heard about this place from the movie *My Girl 2*, which just goes to show that you can learn something about science in the most unexpected places. When I visited the La Brea Tar Pits, I was immediately struck by how central they are. They're right on Wilshire Boulevard, not far from the glamorous shopfronts of Rodeo Drive in

Beverly Hills or the star-studded streets of Hollywood. But the tar pits have been there much, much longer.

Back when mammoths, dire wolves, sabre-toothed cats and giant sloths roamed the California coast, the tar pits were already spitting up their sticky residue to the surface, creating ponds and holes full of natural asphalt. Many animals who were unfortunate enough to set foot in the black slurry got themselves irreversibly trapped. Smaller animals stuck in the tar formed easy prey for larger predators, who then got stuck as well. Thousands of years later, the result is a dense collection of prehistoric fossils just under the surface of one of North America's largest cities.

Palaeontologists are actively exploring the area and you can sometimes see them at work when you visit the Tar Pits and its museum. (In 2019, the Page Museum took on the name of the Tar Pits to make it more clear what they were all about.) One of the park's excavation sites, Pit 91, is currently the longest ongoing urban palaeontological excavation site. It was initially dug between 1913 and 1915 as part of early excavations in that location, but it was revisited in 1969 and has been continuously used as a research site since then. Over the years, scientists have found many extinct plants and animals there that got stuck in the tar tens of thousands of years ago. Pit 91 is now mainly used to study life as it was during the Ice Age, about 25,000 years ago.

The Tar Pits park takes up almost an entire city block. The only other building that shares the same block is the Los Angeles County Museum of Art (LACMA). This museum couldn't be more different from its neighbour. Instead of fossils, it displays art. Instead of dark bubbling puddles of asphalt, its distinguishing outdoor feature is an art installation called "Urban Light", a tidy grid of white historic street lamps that are a popular backdrop for many tourist photos. LACMA has been in its current location, next to the tar pits, since 1965. But the original building has needed some upgrades and renovations over the years. One of these upgrades involved the construction of a new underground parking garage in 2006, and while digging this new space the museum's construction crew came across a mammoth skeleton. Of course, LACMA knew that they could expect to find fossils during this work. After all, the museum was essentially built on top of the tar pits. They still weren't expecting to find nearly an entire mammoth, though!

Around 80% of the mammoth's skeleton was dug out from underneath the future parking garage, including its head and giant tusks. Together with 23 crates of excavated fossils that were also found during this construction phase, the mammoth was sent to the Tar Pits museum next door for research. In the summer of 2008, Tar Pits researchers started work on Project 23, which involved excavating the 23 crates and preparing the mammoth skeleton. Just like previous research at Pit 91, they did a lot of this work in full view of the visiting public. This is how I got to see the mammoth tusks encased in plaster in August of that year, and so did everyone else who visited. It was certainly no secret, but it hadn't been officially announced yet.

Six months later, in February 2009, the Tar Pits museum finally announced to the press what they had been working on for the past months. At that point they had uncovered more than 700 specimens from the parking garage collection, including the mammoth (which they named Zed), but there was a lot more to be done and they would be devoting much of their research to these discoveries. And that's when the rest of the world learned what all the visitors to the Tar Pits had known for many months: that a mammoth was found in a parking garage. But by that time it had already been thousands of years since the animal sealed its fate with an unfortunate misstep into a tar pit, so what's another few months, really?

Manhattanhenge

Hey, there's science in this cityscape

Two times per year, something amazing happens in New York City. Okay, amazing things happen all the time in this big city, but this particular phenomenon doesn't occur more often than that. About three weeks before and after every summer solstice, the setting sun aligns exactly with the city's street grid, causing the low sun to illuminate sides of buildings that would normally be in the shade at that time.

This effect is called "Manhattanhenge", in reference to Stonehenge, an ancient circle of upright stones and stone archways that has stood in the English countryside for thousands of years. Although some people believed that Stonehenge was created by Druids, this is likely not the case at all. The earliest known records that mention the Celtic priests known as Druids are from around Roman times, but Stonehenge is another few thousand years older than that. Scientists who studied DNA from human remains now believe that the people who lived in this area of England at the time that Stonehenge was created were probably Neolithic humans who arrived from other parts of Western Europe.

The area of Stonehenge was in use by humans at least 5,000 years ago (around 3000 BCE), and the stone circle was built a few hundred years after that, around 2500 BCE. The people who built it left no written records about the purpose of Stonehenge, so historians, archaeologists, astronomers and others have had to use other clues to understand what it was for. One clue is that Stonehenge aligns perfectly with the sun on both the summer solstice and winter solstice.

On the summer solstice, the Earth is at such an angle that one of its hemispheres receives the maximum amount of sunlight. For the Northern hemisphere, where Stonehenge is, that falls on or near June 21st. On that day you can see the sun rise exactly behind a large stone placed just outside of the Stonehenge stone circle. This stone, the heel stone, was placed there quite deliberately according to archaeologists. And the most likely purpose is that it had something to do with the solstice. However, experts now believe that the winter solstice might be the day that was most relevant to ancient Stonehenge. On this day the sun also aligns with the stones at Stonehenge, and researchers have found evidence that ancient people held celebrations in this area in winter. What kind of celebrations and why, that remains a mystery, but it's a reasonable bet that it had something to do with the winter solstice.

While Stonehenge is believed to have been built specifically with the solstices in mind, Manhattanhenge is just an artefact of the way that the street grid in this part of New York City is laid out. Manhattan's numbered streets run almost exactly East–West, which is why the Manhattanhenge effect occurs only a few weeks from the solstice, when the Sun reaches its highest point in the sky in summer. During this time of year, sunset is quite late, so the Manhattanhenge effect is usually visible between 8:15 PM and 8:30 PM. The sun then shines directly along the length of Manhattan's streets, lighting buildings and traffic from an unusually low angle.

Because this kind of city grid layout with long uninterrupted roads is quite popular in North America, Manhattan isn't the only place with a Stonehenge-like phenomenon. Any flat city where streets are laid out approximately along a long straight East–West line will do. The further the grid is from the East–West orientation, the further the dates will be from the actual solstices. In Toronto, a similar phenomenon occurs on dates around the winter solstice, in late October and in February. In 2014, Canada's Global News collected eleven years of traffic data from the city and found that the February date of Torontohenge is often one of the peak days for traffic accidents. They suspected that this was caused by the sun blinding drivers on their way home from work, because the sun sets much earlier in February so that Torontohenge coincides with rush hour commute on that day.

Other accidental urban henges, with dates much closer to the actual solstices, can be found, for example, in Washington DC and Baltimore, where the streets run almost exactly along an East–West line. It's not just city streets either. MIT has a very long straight corridor in one of its university buildings, nicknamed the infinite corridor. On a few days in January and November each year, the sunlight passes right through the entire length of the corridor. Predictably, this phenomenon was given the name MIThenge.

Of course, city streets or university buildings were not built with sun alignment in mind. They just happened to align by accident. So how do we know if Stonehenge was purposely built to align with the sun or if that was a coincidence as well? The truth is that we don't really know for sure, but there are some explanations for which there is more evidence than others. Historians' most likely interpretation, based on all the information they have, is that the placement of the stones was done on purpose so that it would align with the sun on solstices, but *why* that was done is not quite understood yet. Stonehenge probably had multiple different roles over the years and people would have visited it for different reasons. So it's maybe not that different from a city after all.

Viral internet meme

Hey, there's science in this viral animation

In early 2016, a science animation went viral on social media. It was a short clip, showing a digital animation of a long lumpy molecule that took tiny steps with its little legs while carrying a huge blobby ball on its back. The clip was widely shared online as an animated GIF, and it was usually accompanied by some text. The text was separate from the animation itself, and at face value it looked like an explanation of what was happening in the clip. According to this text, the molecule was a myosin molecule carrying endorphins, or "happiness molecules", in the brain. It's no wonder that this was a popular clip, widely shared online. Imagine being able to see how molecules make us feel happy!

But as soon as I saw this animation, I knew that the text above it was too good to be true. This clip was similar to images I'd seen many times before in biochemistry textbooks and videos, and from what I could tell, it didn't show a "happiness molecule". It wasn't even a myosin molecule, and it could just as well be the brain as any other organ. So what was in the clip, if not the scientific secret to happiness?

The walking molecule wasn't myosin, but its close cousin kinesin (pronounced kin-NESS-in). Myosin and kinesin are easy to mix up. They're both motor proteins that turn energy into motion. There are even some myosin molecules that carry their cargo just like the molecule in the animation does, such as myosin Va (pronounced five A), but they still look a bit different from kinesin. The biggest clue is that myosin Va has longer "legs", while kinesin has its "feet" very close together, giving it a sort of penguin walk.

The animation also didn't really show a molecule "in the brain". To understand why not, you first need to know that this clip was not taken with a microscope but was a digital representation created by an artist. It shows kinesin in the context of a cell, but the insides of most cells look very similar, so there's no easy way to see whether the animation was set inside a brain cell or in any other type of cell.

And finally, the cargo that our molecular friend is so proudly carrying on its back is not necessarily a "bag of endorphins". Many different molecules are transported in bubble-like compartments such as the one shown in the animation. These compartments are called vesicles, and they're often used to transport molecules towards the outer edge of the cell so that they can be further transported outside of the cell's environment. A vesicle like that *could* be carrying endorphins but it could just as easily be plenty of other things. There's just no way to tell, since the artist only illustrated the outside of the vesicle, and the chemicals inside would be too small to see at this magnification.

Motor proteins are amazing and I loved that a cell biology animation went viral online, but it was unfortunate that the information that someone added to it just wasn't correct, especially when the truth is just as interesting. And as often happens when an image or video goes viral, the artist who created the original didn't get any credit. The original animation that the viral clip was taken from was created by science illustrator and animator John Liebler. He first animated a kinesin molecule for a 2006 video called The Inner Life of the Cell, created together with Harvard researchers. On his website Art of the Cell, Liebler mentions that kinesin wasn't part of the original storyboard for the video at all, but that he decided to create the animation after being inspired by an earlier kinesin animation by another scientific animator, Graham Johnson. Of course, who could resist including a molecule that looks like it takes dapper little steps! It's no wonder it went viral. It might not be a "happiness molecule" or even "in the brain", but looking at that little kinesin molecule stepping along just makes people happy.

Hiking trail geology

Hey, there's science along this long walk

I love hiking on hilly forest trails, where every turn holds a surprise and where a two-hour uphill battle gets rewarded with an amazing view. There aren't many of these places in the parts of Europe where I've spent most of my life, but North America has plenty of them and I love hiking there. During some of the walks I've taken in Algonquin Park, Ontario, or along parts of the Appalachian Trail in the US, one of the things I noticed was that these places carry such clear signs of how the Earth has shifted over time.

When I lived in Toronto, I used to love driving up to Algonquin Park and walking some of the trails there. My favourite was the Lookout Trail, a relatively short two kilometre walk that goes to one of the highest cliffs in the park with views of up to 25 kilometres in the distance. The entire area that you see from the top of the cliff was once much higher ground, but erosion has changed the surface over the years. Signs of a changing landscape start even earlier along that trail, where you come across an enormous rock that was dislodged when the last glacier melted away from the area about 11,000 years ago. The rock was picked up by the glacier and carried for several kilometres before being dropped at the start of what would one day become a hiking trail. In fact, the entire area of Algonquin Park was created by shifts in the

Earth's crust a few hundred million years ago. Because of this, it's now more than 100 metres higher than the surrounding areas. When the crust was lifted up, it cracked in some places, causing cliffs and valleys to appear that now form part of the interesting landscape.

When I left Canada I thought I'd have to miss the experience of hiking in such interesting geological landscapes, but a few years later I had a chance to spend a few days close to the Appalachian Trail on a trip to the United States. The Appalachian Trail is a hiking trail of 2200 miles (more than 3500 kilometres) that runs from Springer Mountain in Georgia to Mount Katahdin in Maine. The trail was created in 1937, so for almost a century people have been able to hike through 14 states along this extremely long trail.

I first heard about the Appalachian Trail when I read Bill Bryson's *A Walk in the Woods*, in which he describes in hilarious detail his experience of hiking the trail with his friend Stephen Katz. Bryson, like many other hikers, attempted to walk the entire distance from end to end, which takes several months. I had no such ambition and certainly not enough time, so I only did a few day hikes.

The first hike was a short one. I found Pine Grove Furnace State Park early in the morning and parked in the almost empty parking lot at the base of Pole Steeple Trail. The only other visitor was a birdwatcher, sitting quietly on a bench overlooking the lake. The main challenge of the short hike up to the overlook point was the sheer number of gnats, but once I reached the cliff and walked out of the shadows of the trees, the gnats suddenly disappeared and I stood in bright sunlight on a rock overlooking Cumberland Valley.

This hilly landscape is part of one of the oldest mountain ridges on Earth and the geological history here is much older than that of Canada's Algonquin Park, which only took shape after the Ice Age. That's just a baby park compared to the Appalachian Mountains, which formed more than 400 million years ago, when all the continents were connected as Pangaea. Most of the prehistoric Central Pangaean Mountains became what is now the Appalachians, but other parts of it turned into the Scottish Highlands and the Atlas mountains in Morocco after the continents drifted apart. After many millions of years, the smooth green hills of Pennsylvania don't look like a rugged mountainous landscape at all any more. Erosion has washed away most of the mountains, so now they're relatively easy to walk, especially for day hikers like me. But the area kept some souvenirs from its more mountainous history. Some rocks were more resistant to erosion than the surrounding ground, so while most of the mountains wore down to hills, the occasional spire of rock remained. That includes the Pole Steeple that this trail was named after, which sticks up above the hills, and formed the cliff from which I looked out over the valley.

Back at the base of the Pole Steeple Trail, I drove further into Pine Grove Furnace State Park. It marks the official middle of the Appalachian Trail, but it was eerily quiet. The Appalachian Trail Museum was closed because it was a Monday in autumn. The Pine Grove Furnace General Store, where in the summer season Appalachian Trail through-hikers are challenged to finish half a gallon of ice cream in one sitting, was also shut for the day. In *A Walk In The Woods*, Bill Bryson encountered a similarly empty park when he passed through here, which made me wonder if it's ever truly busy here.

As the name suggests, Pine Grove Furnace State Park is also home to a furnace. In the 18th and 19th centuries, the park was the location of the township of Pine Grove, built to support the Pine Grove Iron Works. Before transportation of goods was as convenient as it is now, the only way to efficiently deal with iron ore was by processing it on site, even if that meant building an entire settlement in the mountains. Back then, everyone in Pine Grove was in some way involved with the iron industry. Now, all that is left are some old buildings (such as the Ironmaster's Mansion, which is rented out for weddings and parties) and the furnace stack of the old iron works.

I got back in the car and drove to my second hike of the day, at South Mountain State Park. Just across the border in Maryland, this hike took me along a section of the Appalachian Trail to the top of Annapolis Rock. From here I had a beautiful view of the surrounding mountains. On the way up, I came across many hikers who were on their way down the mountain, but when I walked back, I seemed to be all by myself. I heard a mysterious rustling in the bushes a few times. Every time I stopped to look, the rustling stopped. Then I saw it: a chipmunk. And another chipmunk. And a squirrel. And suddenly, the loud cracking of broken branches when two deer darted out of the forest! I was hoping to see some wildlife but wasn't expecting so much of it all at once. And those were just the animals. Along the trail I saw plants, flowers, mushrooms and different kinds of trees. If I had known about it ahead of time, I could have signed up for AT Seasons, a citizen science project that relies on Appalachian Trail hikers to record the wildlife they see along the way. But I just took some photos, got back into the car and drove on to Harpers Ferry.

At the intersection of three states – Virginia, West Virginia and Maryland – Harpers Ferry played an important role in the history of the American Civil War. It's also the location of the headquarters of the Appalachian Trail Conservancy, and unlike the trail museum in Pine Grove, their office was open and welcoming visitors. This, too, was a place I had read about in *A Walk in the Woods*. Visiting in 1997, Bryson found "a large-scale model of the entire trail". And there it was, as promised. It clearly showed how varied the mountains are across the full length of the Appalachians. They were barely hills where I was standing, but they turned into proper mountains further to the South!

Over three separate hikes, I walked a total of only about 16 kilometres of the Appalachian Trail and some of its side trails. That's just 0.5% of the length of the trail, and I already saw so much. But as became very clear from Bryson's writing, the vast majority of the trail is just a very, very long walk in the woods, with little change of scenery and few points of interest. Like Bryson does in the second half of his book, I deliberately drove only to the sections I wanted to explore. After all, you don't have to walk the entire trail for months at a time to get a sense of the impressive geology.

Songs about science

Hey, there's science in these songs

There are songs about everything, so it's no surprise that there are also songs about science. But once you start listening to a lot of these songs, you can't help but notice that they sometimes sound a bit silly. Science songs painstakingly try to fit long words in the lyrics so that they can be scientifically accurate. They're often reductionist or very factually descriptive, and many of them are simply parodies of existing songs in which the words have been changed to something about science.

Despite that, some science songs are really popular. Take for example Tom Lehrer's song "The Elements", set to the tune of Gilbert and Sullivan's "I Am the Very Model of a Modern Major-General". This has become a beloved classic for chemistry fans and a well-known introduction to the periodic table for many people. But it's also very silly. It just lists all the chemical elements.

So why are there so few *serious* songs about science? One reason could be that scientific research is not a familiar daily experience for many people, so it doesn't tend to end up in mainstream music. That could explain why there are more songs about technology than science. After all, email, phones and computers are all deeply integrated into people's everyday lives. For example, Arcade Fire's "We Used To Wait" describes how the experience of waiting for letters to arrive has been displaced by technology.

For many people, scientific research doesn't have as much of a personal connection as applied technology does. Quite the opposite. Science is usually taught as an impersonal series of facts and formulas to describe the world. This impersonal attitude is essential to make science as objective as possible, because the outcomes of scientific research should not depend on who does the research. But thinking of science that way doesn't leave much room for personal opinions or feelings that could be put into song.

Of course, science can't be *that* impersonal because discoveries are made by people, and people are inherently messy and dramatic. So when "scientists" become the topic of songs, rather than science itself, we're suddenly in much more familiar and relatable territory. Several songs about researchers have even made it onto radio stations and award-winning albums. Even these songs still often paint a very stereotypical image of stubborn and headstrong scientists, though. In The Flaming Lips' "Race for the Prize", two scientists are "locked in heated battle", "so determined" to find a cure "if it kills them". In Coldplay's "The Scientist", the metaphorical scientist character understands science better than love. Here, science is the opposite of emotion.

If even songs about people who do science are impersonal, you can imagine how much worse it is for songs that cover science facts and science topics. They're often cheery folk tunes or parodies of pop songs in which words like photosynthesis and deoxyribose are clumsily squeezed into the metre, and where being factual and correct is more important than creating something that's pleasant to listen to. Metaphors are rare. Everything is literal and accurate, and that's what often makes these songs sound silly. They're contrived.

Science songs have such a reputation for silliness that people even joke about it. When country singer Brad Paisley was a guest on TV show

A Prairie Home Companion in 2014, he sang a song, "Geology", about the geological features of Tennessee. The audience laughed before he even started playing, because apparently the entire concept of a song about science was inherently silly. Who would want to listen to a bunch of facts set to music? Still, the comments on the YouTube video of Paisley's performance are full of praise from geologists and geology fans who absolutely love the song. And that perhaps tells you who the audience is for most science songs.

YouTube is full of science songs, but most of them are not originals like Paisley's. Instead, many science songs are parodies in which scientists or science communicators have taken a well-known song and changed the lyrics to be about science. They're often created specifically for science students, scientists or anyone else who loves science a lot. A lot of people who watch these videos know both the original song and the scientific references, so they enjoy seeing how they've been merged in a funny way. It's basically fan art about science and, in that sense, it's very similar to a music style called filk.

Probably the only music genre to get its name through a typo, filk is music created by science fiction fans, about the science fiction universes and characters they love. It originated at sci-fi conventions where people brought their music instruments and sang songs about their favourite fictional worlds based on known melodies. You see the similarity with science songs. Change the topic from science fiction to science, and you end up with popular YouTube science songs like "Bad Project" (a parody of Lady Gaga's "Bad Romance" turned into a song about a failing PhD research project), "Defining Gravity" (a song about Newton's discovery of gravity set to the tune of "Defying Gravity" from the musical *Wicked*) or even the aforementioned "The Elements". Some of these songs are recorded in a lab, others are professional productions, but all these examples are songs by people who love science and like to sing about it.

Much like filk, science songs are usually meant to be shared among people who are in on the joke. People who hear the songs laugh in recognition because they know what it's like to work in a lab, or how many long and difficult names are in the periodic table of elements. Another use for science songs is in education, where the songs serve as a tool to help students learn, so they're crammed full of facts. (The songs, that is. But to be honest, the students are crammed full of facts as well!) If you only consider "science songs" to be these literal and factual songs that fit the filk description, then they are indeed often humorous parodies of existing songs and full of inside jokes and jargon. But there are other songs that allude to science. They aren't always literal, they might be about people instead of facts, they only vaguely hint at scientific concepts, and they are original compositions rather than parodies. They are inspired by science, but they aren't meant to be funny.

In 2016, Kit Buchan put together a list of some great songs inspired by science for *The Observer*. The list includes "Emily" by Joanna Newsom, "Cloudbusting" by Kate Bush, "Higgs Boson Blues" by Nick Cave and the Bad Seeds, "Jupiter Crash" by The Cure, and several others. They're proper songs that just happen to be about science or scientists. You might even have heard some of them before and never even considered that you were listening to a song about science. So yes, lots of science songs are silly, but maybe that's because we only consider the silly ones to be "science songs", and we think of the others as just being regular music.

Painting with egg yolk

Hey, there's science in this paint

For thousands of years, artists have been using a surprising ingredient in some of their paints: egg yolk. This type of paint is called egg tempera and art historians have found it in ancient Egyptian mummy portraits, medieval Italian panel paintings and many other art from around the world. The benefit of this egg yolk-based paint is that it doesn't spread as much as other paints and dries very quickly, which makes it perfect for creating intricate details. But despite being part of artists' toolkits for centuries, there's still a lot to learn about the science of painting with egg yolk.

One of the mysteries of egg tempera is why it maintains its colour so well. Even in paintings that are several centuries old, the pigments have largely kept their original colours if the artists mixed them with egg yolk before applying them to the canvas. In a research study from 2021, chemists at the University of Jaén in Spain exposed different paint samples to UV light to see how the paints would respond. One set of their test samples was a series of pigments prepared in a mixture of egg yolk and linseed oil – a classic variant on a basic egg tempera recipe. After treating these samples with UV radiation, the researchers noticed a few changes in the lipids of the egg yolk, but overall most of the colours in the egg tempera samples were unaffected. It's not clear what makes egg tempera so successful at keeping its colour, but understanding this could make it easier for art conservationists to preserve paintings that are made with this technique.

Another aspect of the egg tempera method that scientists haven't quite figured out is the physics that makes the paint so stretchy. Mixing pigments with egg yolk and another component (like water or oil) produces a very elastic paint that has a very different texture from other types of paints. In 2022, researchers from Sorbonne University in Paris described some of the physical and chemical characteristics of egg tempera. This gave them some new clues about the stretchiness of this paint. In their research, they used a tempera made of green earth pigments and showed that these pigments combine with the proteins in the egg yolk to form a flexible network structure. This is very different from the much looser interaction that you'd normally find between pigment and water molecules in paint, so it could explain why egg tempera paint is so stretchy.

In both research studies, the scientists made their own egg tempera, which is very easy to do – even if you're not a renaissance artist or research chemist. There are several egg tempera recipes online. For example, the Royal Academy of Arts recommends adding five teaspoons of water to an egg yolk and mixing this with any dry pigments you have around the house, such as chalk, make-up or spices. That's similar to what the egg tempera researchers did, and the same technique that artists have been using for centuries to create their paints.

Studying the physics of egg tempera is useful for art historians or art restoration experts who need to preserve old masterpieces that used this technique. It explains why the art ages in the way that it does, and it suggests how best to store it or how to repair any damage. But even if you're not personally in charge of protecting museum art, it's also just interesting to learn why eggs have been used in paintings for so long. Who knew your breakfast could make beautiful art!

Eco homes from earth and mushrooms

Hey, there's science in building with garbage

In many countries, houses are usually built with bricks or wood. But the processes of firing bricks, cutting down trees and transporting building materials are not always the most environmentally friendly ways to build. That's why some people have been exploring new construction methods to make homes more sustainable, and that has led to some unique structures inspired by science, nature, art and ... garbage.

One of my first encounters with eco-friendly buildings was the film *Garbage Warrior*. I saw it at a film festival in 2007 and it made such an impression on me that three years later, when I was in New Mexico for a scientific conference, I took a detour to Taos to check out some of the most creative houses I had ever seen.

Garbage Warrior tells the story of Michael Reynolds, a renegade architect who developed a way to build sustainable homes out of garbage. It may sound a bit disgusting at first, but the homes are beautiful and the way they are designed, built and developed is both architecturally and scientifically experimental. Like any true experimenters, Reynolds and his team tried some things, failed, and made changes and adaptations until they eventually achieved what they set out to do. Reynolds' early customers didn't always understand the experimental aspect of the homes and complained when things didn't work, but over time the buildings have become better and better.

These houses are called Earthships, and the way that they incorporate the local surroundings into the home makes them look a bit like hobbit homes. Usually one side of the building is entirely covered, as if it's shoved into a hill, while the other side has large windows. The orientation of the house is determined by the sun, so that it's at the best angle to be not too hot and not too cold. I stayed in an Earthship for one night when I visited Reynolds' site in Taos and was amazed at how cool it got inside, even in the hot New Mexico summer.

Earthships also use garbage and otherwise unusable materials for structure, insulation and decoration. The outer walls are made of old tyres, filled with mud and cans. Empty glass bottles are arranged in colourful patterns in the walls to let light in and decorate the space. Recycled cans are often used as structural and decorative elements in the kitchen. Each Earthship is unique, because it incorporates materials that the builders have access to.

While Earthships are created as individual homes that incorporate aspects of the location, other designers are trying to create environmentally friendly building materials that can be used at a larger scale in conventional homes. Instead of soil, old bottles and tyres, they're using fungi. To be more specific, they use mycelium, which is a network of fungus fibres that usually grows underground and can sprout mushrooms above ground. Because it grows so quickly and easily, mycelium can be used to sustainably create all kinds of materials, including packaging material, fabrics, bricks, insulation and more.

Using mycelium as a structural and practical material is still a very new process. Even though it's quick to grow and relatively strong for its weight, it can't yet carry the same load as regular bricks can. Mycelium squishes too easily and is still a bit too unpredictable to be used as a structural support in buildings. Researchers are still figuring out how to make the material stronger or how to better control and fine-tune mycelium production so that it can be scaled up, but at the same time, designers and artists are already exploring new ways to use mycelium.

In 2021, one of the honorary mentions of the European Commission's S+T+ARTS Prize (for Innovation in Technology, Industry and Society stimulated by the Arts) prominently featured mycelium. In this project, "The Growing Pavilion", Dutch designers used mycelium as one of the materials in a prototype home that they created entirely without fossil resources. The home demonstrated how various natural materials can be grown and used to create structures and features around the house. The artists used mycelium to create the walls, providing both insulation and decorative elements.

Another mycelium-based building project is MY-CO SPACE by the Berlin-based MY-CO-X collective of artists, designers and biotechnologists. This construction project used mycelium bricks in combination with plywood to create a space that people could enter and explore.

Projects like the Growing Pavilion or MY-CO SPACE are meant to demonstrate some of the possibilities of mycelium as a building material. They're art installations so they're not intended to be mass-produced, but they can inspire other designers and engineers to create new commercial products using these materials. And there are some good reasons why people might want commercial mycelium products.

Mycelium has a low environmental footprint because it can be produced locally using local organic waste. It also grows quickly, which makes it easy to produce large amounts of it. Because it can be grown on waste, it doesn't use soil that could otherwise be used for agriculture, and it repurposes waste by weaving the mycelium network through it to create an entirely new material. This material can then be shaped and processed (for example, by baking) to prevent the fungus from continuing to grow.

Even though the mass-production process of mycelium materials is a very different process from Earthships, with its glass bottles and old tyres built directly into the walls, both methods ultimately re-use waste materials to create homes. So your future home might just be made of garbage instead of wood and bricks.

Ballet turns

Hey, there's science behind ballet turns

Nothing says ballet like a pirouette. To us non-ballerinas even a single pirouette turn can be tricky, but to experienced ballet dancers that's a piece of cake. They can turn several times in a row without losing their balance. These repeated turns are called *"Fouetté"* or *"à la Seconde"*. There's a slight difference in what the legs do in each of these, but both involve multiple rapid repeated turns. It's inevitable that during these repeated turns the dancers shift a bit to keep their balance, but that they're able to do it at all is a remarkable feat. In 2023, researchers in Switzerland explored how ballet dancers manage to stay perfectly balanced while they carry out so many turns in a row, and they discovered how to make the most of an old ballet trick.

Before we get into that research study, let's think about what makes it difficult to make so many turns in a row. The two main challenges are probably getting dizzy and losing your balance. So what would happen if the ballet turns were performed by a robot? Getting a robot to perfectly carry out a pirouette turn without losing its balance turns out to be a persistent challenge in robotics. Even though modern robots are pretty good at correcting their balance for most motions, pirouette turns are particularly tricky for them. And don't even think of asking a robot to do multiple turns in a row.

What makes human dancers different from robots is that they have a brain, and that brain can be tricked into thinking they're not really turning all that much. Ballet dancers keep their head still and facing forward for most of the turn. This minimizes the turning motion in the brain, which helps to prevent dizziness but also makes it easier to stay in balance. Dancers use a trick called "spotting" to make this even easier.

To spot, the dancer starts by looking at a fixed visual cue, such as a spot on the wall in front of them. They then hold their head still and keep making eye contact with that reference point while they start the turn. Once they've turned so far that their head can't stay in that forward-facing position any longer, they quickly whip their head around in the direction of the turn and immediately face forward again to focus on the same spot while their body finishes the turn. All this happens very quickly, and for a sequence of repeated pirouettes it needs to be done at every turn.

But how important is it to have a clear visual reference point when spotting? Can't dancers just keep their head forward as long as possible without keeping their eyes on a fixed spot? That's what researchers at the University of Bern tried to find out. They recruited twelve expert dancers (eight men and four women) and asked them to perform a series of repeated turns. Traditionally, the way these turns are performed can be slightly different in male and female ballet roles, so they looked at two combinations of turns to make sure they covered both ways. The men all did twelve *à la Seconde* turns followed by a double pirouette, and the women started with a double pirouette followed by twelve *Fouetté* turns. All of them first did this exercise in front of a white wall with no obvious visual reference points that they could use for spotting. They were then asked to do it again, but this time a camera was filming them from the front. Unbeknownst to the dancers, this camera wasn't just there to film their turns, but to give the dancers a clear visual reference point that they could use for spotting.

The researchers then compared several measurements between the repeated turns done with or without a reference point and they noticed a difference. When the dancers had a fixed spot in front of them that they could focus their eyes on, such as the camera, they held their balance better and were able to clearly face the front after every turn. The more turns they took, the more important it was to have this visual focus point. The researchers think that this visual point helps the dancers to better create a frame of reference for where their head is, which then helps them to keep the rest of their body in line.

Many dancers will have a favourite point of reference in their usual ballet studios that they use for spotting during practice, but if they're in a new space or on stage during a performance, they'll have to find a new spot for spotting. Knowing that these spots make a real difference, the researchers suggested in their paper that theatres and dance studios should make sure that there is a visual reference point visible from the dance floor or stage to help dancers give their best performances.

It seems so simple that all it takes for ballet dancers to improve their turns is a little spot on the wall for them to focus on. Dancers already knew this, but it's good to know that science has their back.

Can music help you study?

Hey, there's science in your lofi study beats playlist

Some people can't study unless they're listening to music, whether that's a playlist of lofi study beats or their favourite classic vinyl. But other people can only study in silence. Is it just their personal preference or is there something about the music that makes a difference? Over the years, scientists have tried to find the answer in many different ways. They've looked at different types of music, different types of learning, different test subjects, and they've taken different measurements. Some research is focused on discovering whether music affects learning, while other studies try to see if music helps you concentrate. And because these studies all look at slightly different things, some of them find ways in which music is a helpful learning tool while others see music as a distraction. Let's look at some of these research studies, starting with the ones that make the case for having a study playlist.

Learning often involves having to memorize things and later recall them. That's how you would work through lists of vocabulary words, for example, or how you would memorize biology or history facts. In 2018, researchers from the University of Ulm in Germany wanted to see if the way that information is presented makes a difference in how easy it is to memorize. Specifically, they asked whether there was a difference between reading text, hearing spoken wordpro or hearing a song. In this study, researchers Janina Lehmann and Tina Seufert found that it was easiest to memorize text if you read it, but that people who heard it as a song were better able to comprehend the information they learned.

This was good news for educators who use music in the classroom. Some teachers use educational songs in their lessons, where some of the facts that the students need to learn are set to music. These educators noticed that students connect to the material they need to learn in a more engaging way if they sing about it. This method is used in language education, but also for subjects that rely less on memorisation and more on understanding, such as science.

But what if you just prefer to have some background music on while you study? When the music you're listening to isn't relevant to the material that you're studying, your brain is essentially doing two separate tasks: studying and listening to music. Depending on the soundtrack to your study session, the music you're listening to can change your mood. This can make studying easier if you enjoy the music, but it can also distract you from your work. Lehmann and Seufert investigated the role of background music in another study, for which they recruited 81 volunteers (all university students) and made half of them listen to music, while the other half studied in silence.

Originally, they were curious how music would affect memory, but they didn't really find a difference in how well either group was able to memorize what they studied. That suggests that background music didn't have much of an effect on this group. But the researchers did see a correlation between listening to music and being able to understand what you're studying. They noticed that students who had a good working memory found it easier to learn while music was on in the background. The researchers think that for the other students music formed more of a distraction. Essentially, if your brain can handle two separate pieces of information, such as music and study materials, you can study just fine when the music is on and it might make it more enjoyable.

But why do some people prefer to have music on in the background in the first place? One theory is the "arousal-mood hypothesis". This is the idea that listening to fast-paced, upbeat music improves someone's ability of solving a task because it improves their mood. The researchers who first tested this hypothesis used the same Mozart sonata in either a fast or slow tempo, and in a major or minor key, so that all study participants heard the same piece of music but with different characteristics. In this study, people who listened to the faster, major-key version of the piece found it easier to solve a spatial task. A different study also found an effect of happy music on creative problem-solving. So, overall, music seems to help some people study because it changes the way they feel.

In the experiment mentioned above, the researchers chose a Mozart piano piece on purpose because they wanted to control for the so-called "Mozart effect". This is a persistent belief that listening to Mozart's music makes children smarter or makes it easier to learn new things. If only it was so simple. Like many popular theories, there is a tiny grain of truth to it somewhere, but it has been blown out of proportion. In the case of the "Mozart effect", it all dates back to a very short research paper from 1993. This paper describes a study in which 36 university students all either listened to a Mozart piece, to a relaxation tape, or to nothing at all while solving a test. The study didn't include any other music besides Mozart, so the fact that the students who listened to music did a bit better than the others does not say anything about the power of Mozart's music in particular – it just happened to be the music these researchers chose for their music sample. Somehow, these findings took on a life of their own. People read a news article about it, and without reading the original paper, they completely overhyped it to the point where some people believed that listening to Mozart would make babies smarter. (Remember, the study was not even done with babies but with university students.) So, no, listening to Mozart will not make you smarter. It's just that listening to happy, upbeat music may make certain tasks a bit easier.

So that's the case for listening to music while you study. Now let's look at the other side of this. Why do some people say that music distracts them from studying? Remember that one of Lehmann and Seufert's studies found that many people can listen to background music while doing a memory task and several other studies found that listening to happy music can help with creative or spatial tasks. So what should you do if you're not sure what type of studying you're doing? From what we've seen so far, it sounds like you should err on the side of music, because it either has no effect or a positive effect. But I didn't give you the full story yet.

For some people, having music on in the background can make it almost impossible to study. That was something that already came up in Lehmann and Seufert's work, when they noticed that people who didn't have a good working memory would find it hard to learn with music in the background. That's because listening to music uses some of your short-term memory. As you listen to a song, you briefly remember the lyrics you just heard, even if you forget them again seconds later. If you also need to memorize your homework at the same time, you're now doing two tasks that both use your working memory, and that can make it extra difficult.

Several other studies have shown that listening to music can be very distracting in certain learning situations. For example, music is more distracting for introverts or for people who are learning to drive. It also disrupts reading comprehension, especially if the music is fast and loud. Even though upbeat music helps people perform spatial reasoning tasks, as mentioned earlier in this chapter, other studies seem to say that the same kind of music makes it more difficult to do reading comprehension tasks. So, what if you're doing geometry homework, and you need both reading comprehension and spatial reasoning? Music or no music? Why is it so complicated, and why is there no simple answer?

The studies mentioned in this chapter all focused on one very specific aspect of music and learning. Looking at any of these individual studies won't give you the full picture. To get a better overview of what the collective research says about music and learning, I also looked at scientific review articles. These don't just describe one single study, but give an overview of the state of the field. For example, one review article summarizes several studies about the effect of listening to music and found that there is an overall positive effect of "preferred music" on some forms of learning. Another analysis looked specifically at background music and showed that some studies say background music has a positive effect, while others say it's distracting. They did point out that those studies all look at slightly different things, but when they took into account all those different factors, they concluded that "background music disturbs the reading process, has some small detrimental effects on memory, but has a positive impact on emotional reactions and improves achievements in sports."

So there you have it. Listening to music is sometimes distracting, but helpful other times. And whether you're someone who loves a good study beats playlist or someone who needs total silence, you're both right.

The board game Pandemic

Hey, there's science in this board game

Several years before I associated the word "pandemic" with face masks, lockdowns and social distancing, it mainly made me think of a popular board game. The board game Pandemic was first sold in 2008 and rapidly became a favourite of board game fanatics. I first played it at board game cafés but eventually ended up with my own game as well as two of the expansion packs. For many fans of the game, the appeal was in the game design and the collaborative element, but to me it was equally interesting that Pandemic introduced scientific concepts to tabletop gaming.

In Pandemic, all players collaborate to play against the game by trying to cure and treat four rapidly spreading diseases. Each player is assigned a role and each role has a special skill. For example, the dispatcher character can rapidly move other players between cities on the world map that forms the game's playing field. This might seem like a pointless skill, but the ultimate goal of the game is to cure all four diseases, and to cure a disease players need to move their piece to a city on the map that has a research station. They also need to hold multiple cards of the same colour, which can be acquired from other players if they place their pieces in the same city. This way, the dispatcher's power makes it much easier to collect sets of matching colour cards. Other roles in the game include researcher, medic, operations expert and various other jobs. Each profession's special skill is a benefit in the game, but players have to collaborate strategically to get the most out of their roles. For example, while the dispatcher is good at moving players around the board to collect cards, the scientist role makes it even easier by only requiring four cards of the same colour instead of five. If these two roles work together, it suddenly becomes a lot easier for players to defeat the game.

This collaborative combining of specialties is not that different from how researchers work together in real life. Everyone has their own expertise, and together that makes it possible for groups of researchers to do things that they couldn't do on their own. For example, the human genome project, which uncovered all of our DNA code, required people with very different specialties. They needed people with expertise in genetics and others with computational skills, as well as some people who were very detail-oriented and others who were good at seeing the big picture.

Much like research itself, the board game Pandemic also requires coordination and collaboration. If you don't agree on the next few moves with the other players, you won't be able to beat the constantly spreading diseases. And just like in a real pandemic, the worse the infections become, the harder they are to contain. If a city on the map is already heavily infected, this can trigger an outbreak in neighbouring cities, causing a cascade of disease across the board. Players frantically move their pieces around the world map game board to treat imaginary patients while also trying to collect the right cards to collaboratively cure the diseases. It's not always possible to win against the game. If there have been too many outbreaks, or if players run out of playing cards or disease cubes, the game wins.

One of my favourite things about Pandemic is the attention to science-related details. You'll notice the first of them as you set up the game by placing a research centre game piece on the board. The first research centre in the game has to be placed in Atlanta, Georgia, and every player starts the first round here. In real life, Atlanta is home to the Center for Disease Control and Prevention (CDC). You don't need to know this to be able to play the game, but it adds an extra layer of amusement for people who recognize the CDC reference. Another science reference is that in some editions of the game, the game pieces that represent different diseases are stored in little plastic petri dishes which look identical in shape, size and material to those that biologists use in their research labs. I also really appreciated that the images on the role cards look like average biomedical researchers and are suitably diverse.

But in 2020, the game got a little too realistic for me. When everyone was working from home and social distancing to prevent the spread of COVID-19, Pandemic the board game hit too close to home. I didn't need to play a game to be reminded that diseases can spread rapidly around the world and that it takes a while to develop and distribute treatments and vaccines.

That's just me, though. For others, the real pandemic was even more of a reason to take the similarly named board game off the shelf. Board game sales increased in 2020, when people were stuck at home, and Pandemic was one of the titles that many people picked up. Some university professors also used Pandemic in their classrooms to discuss topics like management, collaboration or epidemics. Knowing that people were using this game that I played for fun in a more serious teaching capacity got me to take it off the games shelf again. I didn't want pandemics seeping into my entertainment, but now I saw the game as education. And perhaps I just needed to beat a global pandemic in a game to better cope with the real one. It wouldn't make COVID-19 go away, but at least I had some control over what happened with the fictional diseases on the board. Take that, yellow cubes. We saved the world again.

Acknowledgements

Many of the chapters in this book would not have existed if I hadn't had somewhere to write about science online. So I first want to thank the people who gave me a space to write on the platforms they managed, including Corie Lok, Lou Woodley, Josh Witten, Alex Knapp and other editors and community managers I've worked with over the years.

I also want to thank my science blogging friends in Canada, the US, UK and beyond. You know who you are. I won't list everyone because I don't want to forget anyone, but if we've ever shared a laugh at a science blogging meetup or executed elaborate inside jokes, you're probably responsible for at least some of this book.

For valuable feedback on this collection, I'd like to thank John Calligan and Jo Wimpenny. Thanks also to Talia Pankewycz for checking some of the ballet terms and thanks to Michelle Higgs for proofreading. I also owe many thanks to Sam Jenkins for not only providing feedback on the book itself but having had to deal with me talking about it for over a year.

And finally, thanks to everyone who ever read anything I wrote for the internet in the last (almost) two decades. Your clicks, shares, likes and comments made it all worth it.

Permissions

All chapters in this collection were written by Eva Amsen, but most of them have appeared online in some form prior to this. The author has republication rights for all pieces, either through maintaining rights when publishing on personal or group blogs, or via contractual arrangement with publishers.

Below is a list of all original blog posts and articles that were adapted for this collection, in order of appearance. Each chapter in *Hey, There's Science in This* originated from one or two of the articles below.

"Rubber duckies in research". Copyright © 2008 by Eva Amsen. Reprinted with permission. First published on easternblot.net

"Toys at sea". Copyright © 2014 by Eva Amsen. Reprinted with permission. First published on *The Finch and Pea*

"What makes scary music scary?". Copyright © 2017 by Eva Amsen. Reprinted with permission. First published on Medium.com

"George Washington Carver was an artist long before he became a peanut scientist". Copyright © 2021 by Eva Amsen. Reprinted with permission. First published on Forbes.com

"Yellowstone National Park – birthplace of Taq". Copyright © 2014 by Eva Amsen. Reprinted with permission. First published on *The Finch and Pea*

"Of mice and men – a poem". Copyright © 2015 by Eva Amsen. Reprinted with permission. First published on easternblot.net

"Science of Stranger Things: Dustin's pet research project". Copyright

© 2017 by Eva Amsen. Reprinted with permission. First published on easternblot.net

"Stranger Things And The History And Neuroscience Of Music Therapy". Copyright © 2022 by Eva Amsen. Reprinted with permission. First published on Forbes.com

"Interrailing through Europe with Borodin and Mendeleev". Copyright © 2015 by Eva Amsen. Reprinted with permission. First published on *The Finch and Pea*

"Do computers know what music genres are?" Copyright © 2018 by Eva Amsen. Reprinted with permission. First published on Medium.com

"What the zoo can do for you". Copyright © 2019 by Eva Amsen. Reprinted with permission. First published on easternblot.net

"How A Japanese Anime Boosted Donations To Animal Conservation". Copyright © 2019 by Eva Amsen. Reprinted with permission. First published on Forbes.com

"Dalí's DNA". Copyright © 2006 by Eva Amsen. Reprinted with permission. First published on easternblot.net

"The Vatican Observatory". Copyright © 2014 by Eva Amsen. Reprinted with permission. First published on *The Finch and Pea*

"Seven real life fantastic beasts, and where to find them". Copyright © 2017 by Eva Amsen. Reprinted with permission. First published on easternblot.net

"Distant friends – visiting the School of the Air in Australia". Copyright © 2009 by Eva Amsen. Reprinted with permission. First published on *Expression Patterns* (Nature Network) and republished on easternblot.net

"Mammoth! Or the slowness of science news". Copyright © 2009 by Eva Amsen. Reprinted with permission. First published on easternblot.net

"Pit 91 at the La Brea Tar Pits". Copyright © 2008 by Eva Amsen.

PERMISSIONS

Reprinted with permission. First published on easternblot.net

"Manhattanhenge". Copyright © 2014 by Eva Amsen. Reprinted with permission. First published on *The Finch and Pea*

"Image origins: that walking molecule". Copyright © 2016 by Eva Amsen. Reprinted with permission. First published on easternblot.net

"Algonquin geology". Copyright © 2007 by Eva Amsen. Reprinted with permission. First published on easternblot.net

"Geology and history along the Appalachian Trail". Copyright © 2017 by Eva Amsen. Reprinted with permission. First published on easternblot.net

"Science songs – why so silly?". Copyright © 2016 by Eva Amsen. Reprinted with permission. First published on easternblot.net

"The science of painting with egg yolk". Copyright © 2022 by Eva Amsen. Reprinted with permission. First published on Forbes.com

"Garbage – Part 2: Earthships". Copyright © 2015 by Eva Amsen. Reprinted with permission. First published on *The Finch and Pea*

"The science and art of sustainable mushroom-based building materials". Copyright © 2021 by Eva Amsen. Reprinted with permission. First published on Forbes.com

"Eyes forward: ballet dancers' clever trick to balance their turns". Copyright © 2023 by Eva Amsen. Reprinted with permission. First published on Forbes.com

"Can music help you study?". Copyright © 2018 by Eva Amsen. Reprinted with permission. First published on Medium.com

"When music is a distraction". Copyright © 2018 by Eva Amsen. Reprinted with permission. First published on Medium.com

"Saving the world with Pandemic". Copyright © 2015 by Eva Amsen. Reprinted with permission. First published on easternblot.net

References

Rubber ducks and Lego

- Hohn, D. (2012, February 12). The great escape: The bath toys that swam the Pacific. *The Observer*. https://www.theguardian.com/environment/2012/feb/12/great-escape-bath-toys-pacific
- Ocean motion: Research and applications. Curt Ebbesmeyer. http://oceanmotion.org/html/research/ebbesmeyer.htm
- Ebbesmeyer, C. C., Ingraham, W. J., Royer, T. C., & Grosch, C. E. (2007). Tub toys orbit the pacific subarctic gyre. *Eos, Transactions American Geophysical Union, 88*(1), 1. https://doi.org/10.1029/2007EO010001
- The Cornish beaches where Lego keeps washing up. (2014, July 21). *BBC News*. https://www.bbc.com/news/magazine-28367198
- Associated Press (2003, July 12). Fleet of rubber ducks heads for dry land after 11-year Arctic odyssey. *The Guardian*. https://www.theguardian.com/world/2003/jul/12/arctic
- Can rubber ducks help track a melting glacier? (2008, September 21). *Reuters*. https://www.reuters.com/article/us-climate-glacier-idUSN1746514020080921
- JPL scientist Alberto Behar remembered. NASA Jet Propulsion Laboratory (JPL). https://www.jpl.nasa.gov/news/jpl-scientist-alberto-behar-remembered
- Butler, J. (2022, February 13). 5 million Lego pieces plunged into

the sea off Cornwall. *Cornwall Live.* https://www.cornwalllive.com/news/history/5-million-lego-pieces-plunged-6625201

Scary music

- Halbfinger, D. M. (2005, September 30). His 'secret' movie trailer is no secret anymore. *The New York Times.* https://www.nytimes.com/2005/09/30/movies/his-secret-movie-trailer-is-no-secret-anymore.html
- Rosenbluth, T. (2018, April 12). UCLA professor heads long-term evolutionary biology project on marmots. *Daily Bruin.* https://dailybruin.com/2018/04/12/ucla-professor-heads-long-term-evolutionary-biology-project-on-marmots/
- Blumstein, D. T., Davitian, R., & Kaye, P. D. (2010). Do film soundtracks contain nonlinear analogues to influence emotion? *Biology Letters,* 6(6), 751–754. https://doi.org/10.1098/rsbl.2010.0333
- Burns, J. (2016, March 29). A brief history of the devil's tritone. *Mental Floss.* https://www.mentalfloss.com/article/77321/brief-history-devils-tritone
- Jones, J. (2021, October 4). The medieval ban against the 'devil's tritone': Debunking a great myth in music theory. *Open Culture.* https://www.openculture.com/2021/10/the-medieval-ban-against-the-devils-tritone-debunking-a-great-myth-in-music-theory.html
- Rizzi, S. (2019, March 4) Why did Bernstein build West Side Story around 'the Devil's interval'? *Classic FM.* https://www.classicfm.com/composers/bernstein-l/bernstein-west-side-story-tritone/
- Helmholtz, H. L. F. (1954). *On the Sensations of Tone as a Physiological Basis for the Theory of Music.* New York, NY: Dover Publications.
- Jones, J. (2019, September 19) The creepy 13th-century melody that

shows up in movies again & again: An introduction to 'Dies Irae'. *Open Culture*. https://www.openculture.com/2019/09/the-creepy-13th-century-song-that-shows-up-in-movies-again-again.html

Artist to peanut scientist

- George Washington Carver. Simpson College. https://simpson.edu/dunn-library/george-washington-carver
- The Iowa Pictures (1893, March 14). *The Morning Democrat* (p.4). Davenport, Iowa. https://www.newspapers.com/clip/23234924/george-washington-carvers-painting-on/
- George Washington Carver. American Chemical Society. https://www.acs.org/education/whatischemistry/landmarks/carver.html
- George Washington Carver. Tuskegee University. https://www.tuskegee.edu/support-tu/george-washington-carver
- Reverby, S. M. (2022). The Tuskegee syphilis study. *Oxford Research Encyclopedia of American History*. https://doi.org/10.1093/acrefore/9780199329175.013.1032
- Wright, J. (2021, February 2) More than 'the peanut man'. *U.S. Department of Agriculture.* https://www.usda.gov/media/blog/2014/02/25/more-peanut-man
- Carver peanut products. Tuskegee University. https://www.tuskegee.edu/support-tu/george-washington-carver/carver-peanut-products
- Mackintosh, B. (1977) George Washington Carver and the peanut. *American Heritage.* https://www.americanheritage.com/george-washington-carver-and-peanut
- Art: Black Leonardo. (1941, November 24). *Time*. https://content.time.com/time/subscriber/article/0,33009,801330,00.html

REFERENCES

Lab test's hot spring origins

- Shea, M. (2018, Nov 8) Discovering life in Yellowstone where nobody thought it could exist. *U. S. National Park Service*. https://www.nps.gov/articles/thermophile-yell.htm
- Brock, T. D. (1967). Life at High Temperatures: Evolutionary, ecological, and biochemical significance of organisms living in hot springs is discussed. *Science, 158*(3804), 1012–1019. https://doi.org/10.1126/science.158.3804.1012
- Brock, T. D. (1997). The value of basic research: Discovery of *Thermus aquaticus* and other extreme thermophiles. *Genetics, 146*(4), 1207–1210. https://doi.org/10.1093/genetics/146.4.1207
- Mullis, K. B. (1990). The unusual origin of the polymerase chain reaction. *Scientific American, 262*(4), 56–65. https://www.jstor.org/stable/24996713
- Kossakovski, F. (2021, February 19). The eccentric scientist behind the 'gold standard' COVID-19 test. *National Geographic*. https://www.nationalgeographic.com/science/article/the-eccentric-scientist-behind-the-gold-standard-covid-19-pcr-test
- Mülhardt, C. & Beese, E. W. (2007). The polymerase chain reaction. In *Molecular Biology and Genomics* (pp. 65–94). Elsevier. https://doi.org/10.1016/B978-012088546-6/50006-3
- Gruver, M. (2009, November 30) Parks seek share of profits. *The Spokesman-Review*. https://www.spokesman.com/stories/2009/nov/30/parks-seek-share-of-profits/
- Dawson, P. (2007, November 21). The gold in Yellowstone's microbes. *Time*. https://content.time.com/time/health/article/0,8599,1686700,00.html

Of Mice and Men – a poem

- Nomura, T. (1990). Of mice and men? *Nature, 345*(6277), 671–671. https://doi.org/10.1038/345671a0
- Martin, J. B. (1984). Of mice and men. *Nature, 307*(5946), 10. https://doi.org/10.1038/307010a0
- Danks, D. M. (1986). Of mice and men, metals and mutations. *Journal of Medical Genetics, 23*(2), 99–106. https://doi.org/10.1136/jmg.23.2.99
- Laudenslager, M. L. (2004). Of mice and men, corticosteroids, and vicarious participation. *Brain, Behavior, and Immunity, 18*(5), 414–415. https://doi.org/10.1016/j.bbi.2004.03.001
- Nakamura, T., Takumi, T., Takano, A., Aoyagi, N., Yoshiuchi, K., Struzik, Z. R., & Yamamoto, Y. (2008). Of mice and men – Universality and breakdown of behavioral organization. *PLOS ONE, 3*(4), e2050. https://doi.org/10.1371/journal.pone.0002050
- Romagnani, P. (2013). Of mice and men: The riddle of tubular regeneration: Tubular regeneration in mice and men. *The Journal of Pathology, 229*(5), 641–644. https://doi.org/10.1002/path.4162
- Caplan, A. L. (1980). Of mice and men: The human sciences and the humanities. *The Hastings Center Report, 10*(6), 38. https://doi.org/10.2307/3560300
- Sugarman, J., & Mathews, D. J. H. (2009). Of mice and men: Skin cells, stem cells and ethical uncertainties. *Regenerative Medicine, 4*(6), 791. https://doi.org/10.2217/rme.09.65
- Holliday, R. (1992). Of mice and men. *Nature, 360*(6402), 305–305. https://doi.org/10.1038/360305b0
- Russell, J. (2003). Of mice and men, rats and atherosclerosis. *Cardiovascular Research, 59*(4), 810–811. https://doi.org/10.1016/S0008-6363(03)00530-3
- Szabo, S., Haislip, A. M., & Garry, R. F. (2005). Of mice, cats, and

men: Is human breast cancer a zoonosis? *Microscopy Research and Technique, 68*(3–4), 197–208. https://doi.org/10.1002/jemt.20232
- Ben-Neriah, Y., & Schmitz, M. L. (2004). Of mice and men. *EMBO Reports, 5*(7), 668–673. https://doi.org/10.1038/sj.embor.7400187
- McArthur, J. C., & Kieburtz, K. (2000). Of mice and men: A model of HIV encephalitis. *Neurology, 54*(2), 284–285. https://doi.org/10.1212/wnl.54.2.284
- Borza, D.-B., & Hudson, B. G. (2002). Of mice and men: Murine models of anti-GBM antibody nephritis. *Kidney International, 61*(5), 1905–1906. https://doi.org/10.1046/j.1523-1755.2002.00340.x
- Fuchs, E., & Coulombe, P. A. (1992). Of mice and men: Genetic skin diseases of keratin. *Cell, 69*(6), 899–902. https://doi.org/10.1016/0092-8674(92)90607-e
- Ariyan, S. (1994). Of mice and men – honesty and integrity in medicine: *Annals of Surgery, 220*(6), 745–750. https://doi.org/10.1097/00000658-199412000-00007
- Pennisi, E. (2015). Of mice and men. *Science, 349*(6243), 21–23. https://doi.org/10.1126/science.349.6243.21
- Rew, D. (2003). Of digital mice and men. *European Journal of Surgical Oncology, 29*(7), 624–627. https://doi.org/10.1016/S0748-7983(03)00081-7
- Brown, A. J. (2004). Of cholesterol-free mice and men. *Current Opinion in Lipidology, 15*(3), 373–375. https://doi.org/10.1097/00041433-200406000-00021
- Bateson, M. (2014). Of (Stressed) mice and men. *Nature Methods, 11*(6), 623–624. https://doi.org/10.1038/nmeth.2965
- Leyland-Jones, B., & Grieshaber, C. K. (1999). Of (Only) mice and men. *British Journal of Cancer, 81*(5), 753–755. https://doi.org/10.1038/sj.bjc.6690759
- Kelsh, R., Arnheiter, H., & Bosenberg, M. (Eds.). (2013). Of mice and men ... but so much more too! *Pigment Cell & Melanoma Research,*

26(3), 285–285. https://doi.org/10.1111/pcmr.12102
- Cohen, I. (1991). Of mice, men, and physicians. *The Journal of the Louisiana State Medical Society: Official Organ of the Louisiana State Medical Society, 143*(11), 3–5.
- Of mice, men, and trypanosomes. (1977). *Lancet, 2*(8043), 860–861.
- Aftring, R. P., & Freeman, M. W. (1994). Of mice, men and cholesterol. *Hepatology, 19*(4), 1054–1056. https://doi.org/10.1002/hep.1840190436
- Margolin, J. (2000). Of mice, men, and the genome. *Genome Research, 10*(10), 1431–1432. https://doi.org/10.1101/gr.162800
- Woodruff, A., & Yuste, R. (2008). Of mice and men, and chandeliers. *PLoS Biology, 6*(9), e243. https://doi.org/10.1371/journal.pbio.0060243
- Gordon, M. Y., Lewis, J. L., & Marley, S. B. (2002). Of mice and men ... and elephants. *Blood, 100*(13), 4679–4679. https://doi.org/10.1182/blood-2002-08-2517
- Corballis, M. C. (2008). Of mice and men – and lopsided birds. *Cortex, 44*(1), 3–7. https://doi.org/10.1016/j.cortex.2007.10.001
- Tolkoff-Rubin, N. E. (2002). Of mice and men: The road to tolerance. *Current Opinion in Nephrology and Hypertension, 11*(6), 579–581. https://doi.org/10.1097/00041552-200211000-00001
- Jones, M. E. E., Chin Boon, W., Proietto, J., & Simpson, E. R. (2006). Of mice and men: The evolving phenotype of aromatase deficiency. *Trends in Endocrinology & Metabolism, 17*(2), 55–64. https://doi.org/10.1016/j.tem.2006.01.004
- Berger, A. A. (1991). Of mice and men: An introduction to mouseology or, anal eroticism and Disney. *Journal of Homosexuality, 21*(1–2), 155–166. https://doi.org/10.1300/J082v21n01_11
- Watkin, B. (1978). Of mice, and other beasts, and men. *Nursing Mirror, 146*(4), 10.
- Davies, K. (1993). Of mice and men (And cows and cats). *Nature,*

REFERENCES

361(6411), 478–478. https://doi.org/10.1038/361478a0
- Frank, R. N. (1990). Of mice – And rats, dogs, rabbits, cats, and monkeys – And men. *Archives of Ophthalmology, 108*(4), 502. https://doi.org/10.1001/archopht.1990.01070060050046
- Lewin, B. (1984). Of fruit flies, mice, and men: The illicit review of science. *Cell, 38*(2), 343–344. https://doi.org/10.1016/0092-8674(84)90484-7
- Verity, C. (2000). Of flies, mice, and men. *Developmental Medicine & Child Neurology, 42*(11), 723–723. https://doi.org/10.1111/j.1469-8749.2000.tb00032.x
- Campbell, K. (2003). Of mice and men. *The Lancet Oncology, 4*(6), 334. https://doi.org/10.1016/S1470-2045(03)01088-X
- French, B. A., Yang, Z., Berr, S. S., & Kramer, C. M. (2001). Of mice and men . . . And broken hearts. *Circulation, 104*(21). https://doi.org/10.1161/hc4601.098067
- Hawley, R. G., & Sobieski, D. A. (2002). Of mice and men: The tale of two therapies. *Stem Cells, 20*(4), 275–278. https://doi.org/10.1634/stemcells.20-4-275
- Bennett, V. (1995). Of mice and men: The mice were right. *Journal of Clinical Investigation, 95*(3), 921–922. https://doi.org/10.1172/JCI117797

Science of Stranger Things

- The Pollywog. (2017). *Stranger Things S2E3*. Netflix.
- Dewey Decimal Classification system: commonly used numbers. University of Central Florida Research Guides. https://guides.ucf.edu/dewey/common
- Conant, R. (1984). *Eastern Reptiles and Amphibians*. Easton Press.
- Bevans, M. H. (1956). *The Book of Reptiles and Amphibians*. Doubleday.

- Potenza, A. (2017, November 13). Stranger Things 2 should have had a frog consultant. *The Verge.* https://www.theverge.com/tldr/2017/11/13/16643766/stranger-things-2-pollywog-anachronism-indirana-frog-genus
- Gaitonde, N., & Giri, V. (2014). Primitive breeding in an ancient Indian frog genus Indirana. *Current Science, 107*(1), 109–112. https://www.jstor.org/stable/24103413
- De La Riva, I. (1995). A new reproductive mode for the genus adenomera (Amphibia: Anura: leptodactylidae): Taxonomic implications for certain bolivian and paraguayan populations. *Studies on Neotropical Fauna and Environment, 30*(1), 15–29. https://doi.org/10.1080/01650529509360937
- Busselle, R., & Bilandzic, H. (2008). Fictionality and perceived realism in experiencing stories: A model of narrative comprehension and engagement. *Communication Theory, 18*(2), 255–280. https://doi.org/10.1111/j.1468-2885.2008.00322.x
- Dear Billy. (2022). *Stranger Things S4E4.* Netflix.
- Wheeler, B. L. (2015). *Music Therapy Handbook.* Guilford Publications.
- Heiberger, L., Maurer, C., Amtage, F., Mendez-Balbuena, I., Schulte-Mönting, J., Hepp-Reymond, M.-C., & Kristeva, R. (2011). Impact of a weekly dance class on the functional mobility and on the quality of life of individuals with Parkinson's disease. *Frontiers in Aging Neuroscience, 3*, 14. https://doi.org/10.3389/fnagi.2011.00014
- Bearss, K. A., & DeSouza, J. F. X. (2021). Parkinson's disease motor symptom progression slowed with multisensory dance learning over 3-years: A preliminary longitudinal investigation. *Brain Sciences, 11*(7), 895. https://doi.org/10.3390/brainsci11070895
- Krotinger, A., & Loui, P. (2021). Rhythm and groove as cognitive mechanisms of dance intervention in Parkinson's disease. *PLOS ONE, 16*(5), e0249933. https://doi.org/10.1371/journal.pone.0249

933
- Fischer, C. E., Churchill, N., Leggieri, M., Vuong, V., Tau, M., Fornazzari, L. R., Thaut, M. H., & Schweizer, T. A. (2021). Long-known music exposure effects on brain imaging and cognition in early-stage cognitive decline: A pilot study. *Journal of Alzheimer's Disease, 84*(2), 819–833. https://doi.org/10.3233/JAD-210610
- Gómez Gallego, M., & Gómez García, J. (2017). Music therapy and Alzheimer's disease: Cognitive, psychological, and behavioural effects. *Neurología (English Edition), 32*(5), 300–308. https://doi.org/10.1016/j.nrleng.2015.12.001
- Westbrook, B. K., & McKibben, H. (1989). Dance/movement therapy with groups of outpatients with Parkinson's disease. *American Journal of Dance Therapy, 11*(1), 27–38.
- About the Certification Board for Music Therapists. CBMT. https://www.cbmt.org/about/
- World Federation of Music Therapy. https://wfmt.info/
- Norberg, A., Melin, E., & Asplund, K. (1986). Reactions to music, touch and object presentation in the final stage of dementia. An exploratory study. *International Journal of Nursing Studies, 23*(4), 315–323. https://doi.org/10.1016/0020-7489(86)90054-4
- Odell, H. (1988). A music therapy approach in mental health. *Psychology of Music, 16*(1), 52–61. https://doi.org/10.1177/0305735688161005

A composer among chemists

- Mendeleev, D. *Principles of Chemistry, Vol 2* (1869)
- Baldwin, M. (2019) A look at the first published periodic table. *Physics Today.* https://doi.org/10.1063/PT.6.4.20190327a
- Royal Society of Chemistry. Germanium. https://www.rsc.org/periodic-table/element/32/germanium

- Piano Trio in D Major. Earsense. https://www.earsense.org/chamber-music/Alexander-Borodin-Piano-Trio-in-D-major/
- Mladentsev, M. N. & Tischenko. V. E. *Dmitry Ivanovich Mendeleev: His Life and Work, Vol I.* (1938).
- Dianin, S. *Borodin*, translated by Robert Lord (1963)
- List of works by Aleksandr Borodin. Petrucci Music Library IMSLP. https://imslp.org/wiki/List_of_works_by_Aleksandr_Borodin
- Taylor, J. J. (2016) Musically Russian: Nationalism in the Nineteenth Century. *The Research and Scholarship Symposium.* 4. http://digitalcommons.cedarville.edu/research_scholarship_symposium/2016/podium_presentations/4
- Tony Awards 1954. https://www.tonyawards.com/winners/year/1954/category/any/show/any/
- Figurovskii, N. A. & Solov'ev, Y. I. (1988). Borodin and Education for Women. In: *Aleksandr Porfir'evich Borodin.* Springer, Berlin, Heidelberg. https://doi.org/10.1007/978-3-642-72732-0_11

Music genres

- RobPRocks. (2006) "Pachelbel Rant". https://www.youtube.com/watch?v=JdxkVQy7QLM
- The Axis of Awesome. (2011) "Four Chords". https://www.youtube.com/watch?v=oOlDewpCfZQ
- Pérez-Sancho, C., Rizo, D., Iñesta, J. M., de León, P. J. P., Kersten, S., & Ramirez, R. (2010). Genre classification of music by tonal harmony. *Intelligent Data Analysis, 14*(5), 533–545. https://doi.org/10.3233/IDA-2010-0437
- Huang, P., Wilson, M., Mayfield-Jones, D., Coneva, V., Frank, M., & Chitwood, D. H. (2017). The evolution of Western tonality: a corpus analysis of 24,000 songs from 190 composers over six centuries. *SocArXiv.* https://doi.org/10.31235/osf.io/btshk

- Febres, G., & Jaffe, K. (2017). Music viewed by its entropy content: A novel window for comparative analysis. *arXiv.* https://doi.org/10.48550/arXiv.1510.01806
- Velardo, V. (2019, February 11). Spotify's Discover Weekly explained – Breaking from your music bubble or, maybe not? *The Sound of AI.* https://medium.com/the-sound-of-ai/spotifys-discover-weekly-explained-breaking-from-your-music-bubble-or-maybe-not-b506da144123

The zoo's many roles

- Renner, M. J. (2022, May 23). Where was the world's first zoo? *The Conversation.* http://theconversation.com/where-was-the-worlds-first-zoo-181216
- Miranda, R., Escribano, N., Casas, M., Pino-del-Carpio, A., & Villarroya, A. (2023). The role of zoos and aquariums in a changing world. *Annual Review of Animal Biosciences, 11*(1), 287–306. https://doi.org/10.1146/annurev-animal-050622-104306
- Jensen, E. (2014). Evaluating children's conservation biology learning at the zoo. *Conservation Biology, 28*(4), 1004–1011. https://doi.org/10.1111/cobi.12263
- Moss, A., Jensen, E., & Gusset, M. (2015). Evaluating the contribution of zoos and aquariums to Aichi Biodiversity Target 1. *Conservation Biology, 29*(2), 537–544. https://doi.org/10.1111/cobi.12383
- Fukano, Y., Tanaka, Y., & Soga, M. (2020). Zoos and animated animals increase public interest in and support for threatened animals. *Science of The Total Environment, 704,* 135352. https://doi.org/10.1016/j.scitotenv.2019.135352
- Conde, D. A., Staerk, J., Colchero, F., Da Silva, R., Schöley, J., Baden, H. M., Jouvet, L., Fa, J. E., Syed, H., Jongejans, E., Meiri, S., Gaillard,

- J.-M., Chamberlain, S., Wilcken, J., Jones, O. R., Dahlgren, J. P., Steiner, U. K., Bland, L. M., Gomez-Mestre, I., ... Vaupel, J. W. (2019). Data gaps and opportunities for comparative and conservation biology. *Proceedings of the National Academy of Sciences, 116*(19), 9658–9664. https://doi.org/10.1073/pnas.1816367116
- California condor reintroduction & recovery. U. S. National Park Service. https://www.nps.gov/articles/california-condor-recovery.htm
- Rogers, K. (2015, August 24). Why is it so hard to breed giant pandas? *Vice*. https://www.vice.com/en/article/jp5wax/why-is-it-so-hard-to-breed-giant-pandas
- Jones, B. (2023, January 17). We pulled pandas back from the brink of extinction. Meanwhile, the rest of nature collapsed. *Vox*. https://www.vox.com/the-highlight/23540896/pandas-extinction-conservation-zoos-wwf-climate
- How did China save the giant panda? (2016, September 5). *BBC News*. https://www.bbc.com/news/world-asia-china-37273337

Late night Japanese anime

- Fukano, Y., Tanaka, Y., & Soga, M. (2020). Zoos and animated animals increase public interest in and support for threatened animals. *Science of The Total Environment, 704*, 135352. https://doi.org/10.1016/j.scitotenv.2019.135352
- けものフレンズでにくがnhkにのは (2017, February 15). *Huffington Post Japan*. https://www.huffingtonpost.jp/2017/02/15/kemono-frends-nhk_n_14783606.html
- Kemono Friends. https://kemono-friends-20170110.jp/zoo/
- The IUCN Red List of Threatened Species. https://www.iucnredlist.org/

Dalí and DNA

- Salvador Dalí: The persistence of memory. The Museum of Modern Art. https://www.moma.org/collection/works/79018
- The structure of DNA. Stereoscopic work. Fundació Gala – Salvador Dalí. https://www.salvador-dali.org/en/artwork/catalogue-raisonne-paintings/obra/1107/the-structure-of-dna-stereoscopic-work
- Dalí. Stereoscopic images. Painting in three dimensions. Dalí Exhibitions. https://exhibitions.salvador-dali.org/en/daliestereoscopies/
- Stereo viewing. https://spdbv.unil.ch/TheMolecularLevel/0Help/StereoView.html
- Prosser, K. E., Stokes, R. W., & Cohen, S. M. (2020). Evaluation of 3-dimensionality in approved and experimental drug space. *ACS Medicinal Chemistry Letters*, *11*(6), 1292–1298. https://doi.org/10.1021/acsmedchemlett.0c00121
- PDB101: Learn: guide to understanding PDB data: methods for determining structure. RCSB: PDB-101. https://pdb101.rcsb.org/learn/guide-to-understanding-pdb-data/methods-for-determining-structure
- Brown, W. H. (1995). *Organic Chemistry* (International Edition). Saunders College Publishing.
- When magic eye pictures ruled the world – and frustrated millions of people. (2020, May 7). *Mental Floss*. https://www.mentalfloss.com/article/622658/when-magic-eye-pictures-ruled-world
- Guardiola, E., & Baños, J.-E. (2003). Dalí and the double helix. *Nature*, *423*(6943), 917–917. https://doi.org/10.1038/423917a
- Boyce, J. J. S. (2007, November 28). The art of science – the science of art? *Science Creative Quarterly*. https://www.scq.ubc.ca/the-art-of-science-the-science-of-art/

The Vatican Observatory

- Hattenbach, J. (2023, January 6). Light Pollution Is Creeping Up on the World's Observatories. *Sky & Telescope.* https://skyandtelescope.org/astronomy-news/light-pollution-is-creeping-up-on-worlds-observatories/
- Vatican Observatory. https://www.vaticanobservatory.org/
- Mount Graham International Observatory. https://mgio.arizona.edu/
- Ten days that vanished: The switch to the Gregorian calendar. *Encyclopedia Britannica.* https://www.britannica.com/story/ten-days-that-vanished-the-switch-to-the-gregorian-calendar
- Vatican hosts study week on astrobiology. (2009) *NASA Astrobiology.* https://astrobiology.nasa.gov/news/vatican-hosts-study-week-on-astrobiology/
- Vatican Observatory – Voss 2023. https://www.vaticanobservatory.va/en/education/voss/voss-2023
- Pinkerton, B. (2019, July 19). What would alien life mean for Catholicism? The Vatican's chief astronomer explains. *Vox.* https://www.vox.com/science-and-health/2019/7/19/20698391/vatican-astronomer-guy-consolmagno-meteorites-intelligent-life-galileo-religion
- Vatican admits Galileo was right. (1992, November 7). *New Scientist.* https://www.newscientist.com/article/mg13618460-600-vatican-admits-galileo-was-right/
- History. Vatican Observatory. https://www.vaticanobservatory.va/en/history

A bestiary of amazing animals

- Rochester Cathedral. Manticore (Manticora). https://www.rochestercathedral.org/research/bestiary/manticore
- Basilisk. The Medieval Bestiary. https://bestiary.ca/beasts/beast265.htm
- Ings, S. (2023, November 8). The beauty of medieval bestiaries. *The Spectator.* https://www.spectator.co.uk/article/the-beauty-of-medieval-bestiaries/
- An introduction to the bestiary, book of beasts in the medieval world. Getty. https://www.getty.edu/news/an-introduction-to-the-bestiary-book-of-beasts-in-the-medieval-world
- Impey, O. (1992). The naming of the beasts. Natural history in the medieval bestiary. *Journal of the History of Collections, 4*(1), 156–156. https://doi.org/10.1093/jhc/4.1.156
- Rowling, J. K. (2001). *Fantastic Beasts and Where to Find Them.* Bloomsbury
- Reiß, C., Olsson, L., & Hoßfeld, U. (2015). The history of the oldest self-sustaining laboratory animal: 150 years of axolotl research. *Journal of Experimental Zoology Part B: Molecular and Developmental Evolution, 324*(5), 393–404. https://doi.org/10.1002/jez.b.22617
- Ambystoma mexicanum. IUCN Red List of Threatened Species. https://www.iucnredlist.org/species/1095/53947343
- Platypus. *National Geographic.* https://www.nationalgeographic.com/animals/mammals/facts/platypus
- The coelacanth story. Knysna Museums. https://www.knysnamuseums.co.za/pages/the-coelacanth/
- Amemiya, C. T., Alföldi, J., Lee, A. P., Fan, S., Philippe, H., MacCallum, I., Braasch, I., Manousaki, T., Schneider, I., Rohner, N., Organ, C., Chalopin, D., Smith, J. J., Robinson, M., Dorrington, R. A., Gerdol, M., Aken, B., Biscotti, M. A., Barucca, M., … Lindblad-

- Toh, K. (2013). The African coelacanth genome provides insights into tetrapod evolution. *Nature, 496*(7445), 311–316. https://doi.org/10.1038/nature12027
- Smith, C. L., Rand, C. S., Schaeffer, B., & Atz, J. W. (1975). *Latimeria*, the living coelacanth, is ovoviviparous. *Science, 190*(4219), 1105–1106. https://doi.org/10.1126/science.190.4219.1105
- Fearnley, K. (2016, January 30) Weird & wonderful creatures: The aye-aye. *AAAS*. https://www.aaas.org/news/weird-wonderful-creatures-aye-aye
- Sieger, J., Brümmer, F., Ahn, H., Lee, G., Kim, S., & Schill, R. O. (2022). Reduced ageing in the frozen state in the tardigrade *Milnesium inceptum* (Eutardigrada: Apochela). *Journal of Zoology, 318*(4), 253–259. https://doi.org/10.1111/jzo.13018
- Tsujimoto, M., Imura, S., & Kanda, H. (2016). Recovery and reproduction of an Antarctic tardigrade retrieved from a moss sample frozen for over 30 years. *Cryobiology, 72*(1), 78–81. https://doi.org/10.1016/j.cryobiol.2015.12.003
- Jönsson, K. I. (2019). Radiation tolerance in tardigrades: Current knowledge and potential applications in medicine. *Cancers, 11*(9), 1333. https://doi.org/10.3390/cancers11091333
- Tiny animals survive exposure to space. ESA. https://www.esa.int/Science_Exploration/Human_and_Robotic_Exploration/Research/Tiny_animals_survive_exposure_to_space
- Neves, R. C., Hvidepil, L. K. B., Sørensen-Hygum, T. L., Stuart, R. M., & Møbjerg, N. (2020). Thermotolerance experiments on active and desiccated states of *Ramazzottius varieornatus* emphasize that tardigrades are sensitive to high temperatures. *Scientific Reports, 10*(1), 94. https://doi.org/10.1038/s41598-019-56965-z
- Goldstein, B. (2022). Tardigrades and their emergence as model organisms. *Current Topics in Developmental Biology, 147*, 173–198. https://doi.org/10.1016/bs.ctdb.2021.12.008

REFERENCES

- Anderson, R. C., & Wood, J. B. (2001). Enrichment for giant pacific octopuses: Happy as a clam? *Journal of Applied Animal Welfare Science, 4*(2), 157–168. https://doi.org/10.1207/S15327604JAWS0402_10
- How many hearts does an octopus have? *New Scientist.* https://www.newscientist.com/question/many-hearts-octopus/
- Giant Pacific octopus. (2011, June 10). *National Geographic.* https://www.nationalgeographic.com/animals/invertebrates/facts/giant-pacific-octopus
- Heinrich, S., Wittmann, T. A., Prowse, T. A. A., Ross, J. V., Delean, S., Shepherd, C. R., & Cassey, P. (2016). Where did all the pangolins go? International CITES trade in pangolin species. *Global Ecology and Conservation, 8,* 241–253. https://doi.org/10.1016/j.gecco.2016.09.007

School of the Air

- Alice Springs School of the Air Visitor Centre. https://www.schooloftheair.net.au/
- Interactive and Engaging Distance Education. HundrED. https://hundred.org/en/innovations/interactive-and-engaging-distance-education
- What is a governess? Australian Nanny Association. https://www.australiannannyassociation.org.au/what-is-a-governess/
- Australian school builds community despite vast distances in the Outback. (2012, May 23). *CNN.* https://www.cnn.com/2012/05/23/us/australian-school-builds-community-despite-vast-distances-in-the-outback/index.html
- Cortés-Albornoz, M. C., Ramírez-Guerrero, S., García-Guáqueta, D. P., Vélez-Van-Meerbeke, A., & Talero-Gutiérrez, C. (2023). Effects of remote learning during COVID-19 lockdown on chil-

dren's learning abilities and school performance: A systematic review. *International Journal of Educational Development, 101*, 102835. https://doi.org/10.1016/j.ijedudev.2023.102835
- Dabrowski, A., Nietschke, Y., Taylor-Guy, P., & Chase, A-M. (2020). *Mitigating the impacts of COVID-19: Lessons from Australia in remote education.* Australian Council for Educational Research. https://doi.org/10.37517/978-1-74286-618-5
- Widdicombe, L. (2020, April 2). The great zoom-school experiment. *The New Yorker.* https://www.newyorker.com/news/our-local-correspondents/the-great-zoom-school-experiment
- Golden, A. R., Srisarajivakul, E. N., Hasselle, A. J., Pfund, R. A., & Knox, J. (2023). What was a gap is now a chasm: Remote schooling, the digital divide, and educational inequities resulting from the COVID-19 pandemic. *Current Opinion in Psychology*, 101632. https://doi.org/10.1016/j.copsyc.2023.101632
- Rodell, B. (2020, April 10). Remote learning is nothing new to outback families. *New York Times.* https://www.nytimes.com/2020/04/10/world/australia/australia-remote-learning-schools-coronavirus.html

Museum parking garage

- Wyatt, E. (2009, February 19). Los Angeles tar yields mammoth's skeleton. *New York Times.* https://www.nytimes.com/2009/02/19/us/19labrea.html
- Holden, A. R., & Southon, J. R. (2016). Radiocarbon dating and stable isotopic analysis of insect chitin from the Rancho La Brea Tar Pits, southern California. *Radiocarbon, 58*(1), 99–113. https://doi.org/10.1017/RDC.2015.9
- Holden, A. R., Harris, J. M., & Timm, R. M. (2013). Paleoecological and taphonomic implications of insect-damaged Pleistocene ver-

REFERENCES

- tebrate remains from Rancho La Brea, southern California. *PLOS ONE, 8*(7), e67119. https://doi.org/10.1371/journal.pone.0067119
- History. La Brea Tar Pits. https://tarpits.org/la-brea-tar-pits-history
- Pit 91. La Brea Tar Pits. https://tarpits.org/experience-tar-pits/pit-91
- The story of urban light. (2018, February 6). *Unframed - LACMA*. https://unframed.lacma.org/2018/02/06/story-urban-light
- Project 23. La Brea Tar Pits. https://tarpits.org/experience-tar-pits/project-23
- Maugh, T. H., II (2009, February 18). Major cache of fossils unearthed in L.A. *Los Angeles Times*. https://www.latimes.com/archives/la-xpm-2009-feb-18-sci-fossils18-story.html

Manhattanhenge

- Iezzi, T. (2015, May 29). How Neil deGrasse Tyson Discovered Manhattanhenge. *Fast Company*. https://www.fastcompany.com/3046868/how-neil-degrasse-tyson-discovered-manhattanhenge
- Blakemore, E. (2019, November 18). Why do we know so little about the Druids? *National Geographic*. https://www.nationalgeographic.co.uk/history-and-civilisation/2019/11/why-do-we-know-so-little-about-the-druids
- History of Stonehenge. English Heritage. https://www.english-heritage.org.uk/visit/places/stonehenge/history-and-stories/history/
- Greenspan, R. E. (2019, June 20). Here's why Stonehenge is connected to the summer solstice. *Time*. https://time.com/5608296/summer-solstice-stonehenge-history/
- Madgwick, R., Lamb, A. L., Sloane, H., Nederbragt, A. J., Albarella, U., Pearson, M. P., & Evans, J. A. (2019). Multi-isotope analysis reveals that feasts in the Stonehenge environs and across Wessex

drew people and animals from throughout Britain. *Science Advances*, 5(3), eaau6078. https://doi.org/10.1126/sciadv.aau6078
- Cain, P. (2014, March 7) Sun in your eyes: 'Torontohenge' sees accidents spike – Toronto. *Global News*. https://globalnews.ca/news/1194051/torontohenge-sees-accident-spike/
- Illuminating the Corridor is an annual solar event. (2005, January 26). *MIT News | Massachusetts Institute of Technology*. https://news.mit.edu/2005/iap-mithenge-0126
- Understanding Stonehenge. English Heritage. https://www.english-heritage.org.uk/visit/places/stonehenge/history-and-stories/understanding-stonehenge/

Viral internet meme

- Kasprak, A. (2017, July 26). Is this 'walking' protein what happiness really looks like? *Snopes*. https://www.snopes.com/fact-check/what-happiness-looks-like/
- Goldman, Y. E., & Ostap, E. M. (2012). 4. 9 myosin motors: Kinetics of myosin. In E. H. Egelman (Ed.), *Comprehensive Biophysics* (pp. 151–169). Elsevier. https://doi.org/10.1016/B978-0-12-374920-8.00411-2
- The inner Life of the Cell. XVIVO Scientific Animation. https://xvivo.com/examples/the-inner-life-of-the-cell/
- Liebler, J. (2014, May 9) Well, you can tell by the way I use my walk… Scientific illustration. *Medical & Scientific Video Animation & Illustration* https://www.artofthecell.com/well-you-can-tell-by-the-way-i-use-my-walk/

REFERENCES

Hiking trail geology

- Lookout trail. Algonquin Provincial Park. The friends of Algonquin Park. https://www.algonquinpark.on.ca/visit/recreational_activites/lookout-trail.php
- Guillet, G. R. (1969). *Algonquin Provincial Park. A Geological Guide to Highway 60.* Ontario Department of Mines.
- Appalachian national scenic trail. U. S. National Park Service. https://www.nps.gov/appa/index.htm
- Geologic features. Pennsylvania Department of Conservation & Natural Resources. https://www.dcnr.pa.gov:443/Recreation/WhereToGo/GeologicFeatures/Pages/default.aspx
- Bryson, B. (2007). *A Walk in the Woods.* Anchor. (Original work published 1998)
- Hatcher Jr. , R. D. (2010). The Appalachian orogen: A brief summary. In *From Rodinia to Pangea: The Lithotectonic Record of the Appalachian Region.* Geological Society of America.
- Sholtis, B. (2016, July 6). *Hikers bond over PA's ice cream challenge.* York Daily Record. https://www.ydr.com/story/life/2016/07/06/hikers-bond-over-pas-ice-cream-challenge/85997768/
- History of pine grove furnace state park. Pennsylvania Department of Conservation & Natural Resources. https://www.dcnr.pa.gov:443/StateParks/FindAPark/PineGroveFurnaceStatePark/Pages/History.aspx
- A.T. Seasons Project. US National Park Service. https://www.nps.gov/articles/a-t-seasons-project.htm
- Appalachian Trail Conservancy. https://appalachiantrail.org

Songs about science

- Tom Lehrer (1959). "The Elements" on *More of Tom Lehrer*. Lehrer Records
- Gilbert and Sullivan (1879). "I Am the Very Model of a Modern Major-General". From *The Pirates of Penzance*
- Arcade Fire (2010). "We Used To Wait" on *The Suburbs*. Merge; City Slang; Mercury
- The Flaming Lips (1999). "Race For The Prize" on *The Soft Bulletin*. Warner Records
- Coldplay (2002). "The Scientist" on *A Rush of Blood To The Head*. Parlophone; Capitol Records
- Brad Paisley (2014). "Geology". Performed live at *A Prairie Home Companion* on May 10th, 2014
- Polo, S. (2011, April 12). Filk-ing around: A brief history of geeky, folky, fan music. *The Mary Sue*. https://www.themarysue.com/filk-ing-around-a-brief-history-of-geeky-folky-fan-music/
- Zheng Lab at Baylor College of Medicine (2011). "Bad Project". https://www.youtube.com/watch?v=Fl4L4M8m4d0
- Lady Gaga (2009). "Bad Romance" on *The Fame Monster*. Streamline; KonLive; Cherrytree; Interscope
- A Capella Science featuring Physics Girl and Malinda (2016). "Defining Gravity". https://www.youtube.com/watch?v=yTQyFzTl5GA
- Stephen Schwartz (2003). "Defying Gravity" written for *Wicked*
- Buchan, K. (2016, February 14). The lab playlist: 16 great songs about science (And a bad one). *The Observer*. https://www.theguardian.com/music/2016/feb/14/the-lab-playlist-16-great-songs-about-science-and-a-bad-one

Painting with egg yolk

- González-Cabrera, M., Domínguez-Vidal, A., & Ayora-Cañada, M. J. (2021). Monitoring UV-accelerated alteration processes of paintings by means of hyperspectral micro-FTIR imaging and chemometrics. *Spectrochimica Acta Part A: Molecular and Biomolecular Spectroscopy, 253,* 119568. https://doi.org/10.1016/j.saa.2021.119568
- Fanost, A., De Viguerie, L., Ducouret, G., Mériguet, G., Walter, P., Glanville, H., Rollet, A., & Jaber, M. (2022). Connecting rheological properties and molecular dynamics of egg-tempera paints based on egg yolk. *Angewandte Chemie International Edition, 61*(1), e202112108. https://doi.org/10.1002/anie.202112108
- Smallman, A. (2021, May 25). How to: Make tempera paint. *Royal Academy of Arts.* https://www.royalacademy.org.uk/article/how-to-make-egg-tempera-paint

Eco homes from earth and mushrooms

- Hodge, O. (2007) *Garbage Warrior.* Open Eye Media; ITVS International
- Ip, K., & Miller, A. (2009). Thermal behaviour of an earth-sheltered autonomous building – The Brighton Earthship. *Renewable Energy, 34*(9), 2037–2043. https://doi.org/10.1016/j.renene.2009.02.006
- Xing, Y., Brewer, M., El-Gharabawy, H., Griffith, G., & Jones, P. (2018). Growing and testing mycelium bricks as building insulation materials. *IOP Conference Series: Earth and Environmental Science, 121,* 022032. https://doi.org/10.1088/1755-1315/121/2/022032
- Gou, L., Li, S., Yin, J., Li, T., & Liu, X. (2021). Morphological and physico-mechanical properties of mycelium biocomposites with natural reinforcement particles. *Construction and Building Materials,*

304, 124656. https://doi.org/10.1016/j.conbuildmat.2021.124656
- Antinori, M. E., Ceseracciu, L., Mancini, G., Heredia-Guerrero, J. A., & Athanassiou, A. (2020). Fine-tuning of physicochemical properties and growth dynamics of mycelium-based materials. *ACS Applied Bio Materials, 3*(2), 1044–1051. https://doi.org/10.1021/acsabm.9b01031
- S+T+ARTS Prize: Science, Technology, and the ARTS. https://starts-prize.aec.at/en/
- The Growing Pavilion. S+T+ARTS Prize. https://starts-prize.aec.at/en/the-growing-pavilion/
- MY-CO-SPACE: MY-CO-X. https://tinybe.org/en/artists/my-co-x/

Ballet turns

- Bauer, C. (2020, December 13). What makes these 6 turns so hard – Plus tips on how to master them. *Dance Magazine.* https://www.dancemagazine.com/tricky-ballet-turns/
- Schärli, A., Haber, C., & Klostermann, A. (2023). Does a visual reference help ballet dancers turn more successfully? *Human Movement Science, 88,* 103062. https://doi.org/10.1016/j.humov.2023.103062
- Shiriaev, A. S., Freidovich, L. B., & Manchester, I. R. (2008). Can we make a robot ballerina perform a pirouette? Orbital stabilization of periodic motions of underactuated mechanical systems. *Annual Reviews in Control, 32*(2), 200–211. https://doi.org/10.1016/j.arcontrol.2008.07.001

Can music help you study?

- Lehmann, J. A. M., & Seufert, T. (2018). Can music foster learning – effects of different text modalities on learning and information retrieval. *Frontiers in Psychology, 8.* https://www.frontiersin.org/articles/10.3389/fpsyg.2017.02305
- Ludke, K. M., Ferreira, F., & Overy, K. (2014). Singing can facilitate foreign language learning. *Memory & Cognition, 42*(1), 41–52. https://doi.org/10.3758/s13421-013-0342-5
- Science songs. Songs for Teaching. https://www.songsforteaching.com/sciencesongs.htm
- Lehmann, J. A. M., & Seufert, T. (2017). The influence of background music on learning in the light of different theoretical perspectives and the role of working memory capacity. *Frontiers in Psychology, 8.* https://www.frontiersin.org/articles/10.3389/fpsyg.2017.01902
- Husain, G., Thompson, W. F., & Schellenberg, E. G. (2002). Effects of musical tempo and mode on arousal, mood, and spatial abilities. *Music Perception, 20*(2), 151–171. https://doi.org/10.1525/mp.2002.20.2.151
- Ritter, S. M., & Ferguson, S. (2017). Happy creativity: Listening to happy music facilitates divergent thinking. *PLOS ONE, 12*(9), e0182210. https://doi.org/10.1371/journal.pone.0182210
- Rauscher, F. H., Shaw, G. L., & Ky, C. N. (1993). Music and spatial task performance. *Nature, 365*(6447), 611–611. https://doi.org/10.1038/365611a0
- Pietschnig, J., Voracek, M., & Formann, A. K. (2010). Mozart effect–Shmozart effect: A meta-analysis. *Intelligence, 38*(3), 314–323. https://doi.org/10.1016/j.intell.2010.03.001
- McCord, M. A., & Joseph, D. L. (2020). A framework of negative responses to introversion at work. *Personality and Individual*

Differences, *161*, 109944. https://doi.org/10.1016/j.paid.2020.1 09944
- Brodsky, W., & Slor, Z. (2013). Background music as a risk factor for distraction among young-novice drivers. *Accident Analysis & Prevention*, *59*, 382–393. https://doi.org/10.1016/j.aap.2013.06.02 2
- Thompson, W. F., Schellenberg, E. G., & Letnic, A. K. (2012). Fast and loud background music disrupts reading comprehension. *Psychology of Music*, *40*(6), 700–708. https://doi.org/10.1177/0305 735611400173
- Pauwels, E. K. J., Volterrani, D., Mariani, G., & Kostkiewics, M. (2014). Mozart, music and medicine. *Medical Principles and Practice*, *23*(5), 403–412. https://doi.org/10.1159/000364873
- Kämpfe, J., Sedlmeier, P., & Renkewitz, F. (2011). The impact of background music on adult listeners: A meta-analysis. *Psychology of Music*, *39*(4), 424–448. https://doi.org/10.1177/030573561037 6261

The board game Pandemic

- Pandemic. BoardGameGeek. https://boardgamegeek.com/boardg ame/30549/pandemic
- Vermeulen, N., Parker, J. N., & Penders, B. (2013). Understanding life together: A brief history of collaboration in biology. *Endeavour*, *37*(3), 162–171. https://doi.org/10.1016/j.endeavour.2013.03.001
- Beattie, S. (2020, April 29). Playing Pandemic – The hit board game about the very thing we're trying to avoid. *The Conversation*. http://theconversation.com/playing-pandemic-the-hit-board-ga me-about-the-very-thing-were-trying-to-avoid-137009
- Van Esch, C., & Wiggen, T. (2020). Can your students save the world? Utilizing *Pandemic®*, a cooperative board game, to

teach management. *Management Teaching Review, 5*(3), 275–283. https://doi.org/10.1177/2379298120933626
- Anania, E. C., Keebler, J. R., Anglin, K. M., & Kring, J. P. (2016). Using the cooperative board game *Pandemic* to study teamwork. *Proceedings of the Human Factors and Ergonomics Society Annual Meeting, 60*(1), 1770–1774. https://doi.org/10.1177/1541931213601405
- MacIntyre, G. (2020, September 10) Pandemic, the class: Real-time simulation-based course challenges students to prevent virus outbreaks. *Dalhousie News.*

(All URLs accessed February 13th, 2024)

About the Author

Eva Amsen is a science writer based in London, UK. She has written about science and scientists for several publications, including *Nature*, the *British Medical Journal*, *Undark*, *Hakai*, and the Forbes website. Eva got her start as a writer by running a science blog while she was supposed to be focusing on her Biochemistry PhD at the University of Toronto. She's still proud of some of the pieces she wrote for a handful of readers on the internet back in the day and is excited to reintroduce you to some of her old writing, as well as to newer pieces, in *Hey, There's Science In This*.

You can connect with me on:
- https://evaamsen.com